Power Sy

Voltage Stability

BRIAN GLOVER
MAPP COR
1125 ENERGY PARK DR.
ST PAUL MN 55108
651 632 8528
bp.glover@mapp.org

get the errata sheet

The EPRI Power System Engineering Series
Dr. Neal J. Balu, Editor-in-Chief

KUNDUR • *Power System Stability and Control*
TAYLOR • *Power System Voltage Stability*

Power System Voltage Stability

Carson W. Taylor
Fellow IEEE
Principal, Carson Taylor Seminars
Principal Engineer, Bonneville Power Administration

Edited by

Neal J. Balu
Dominic Maratukulam
Power System Planning and Operations Program

Electrical Systems Division
Electric Power Research Institute
3412 Hillview Avenue
Palo Alto, California

New York San Francisco Washington, D.C. Auckland Bogotá
Caracas Lisbon London Madrid Mexico City Milan
Montreal New Delhi San Juan Singapore
Sydney Tokyo Toronto

POWER SYSTEM VOLTAGE STABILITY
International Editions 1994

10 09 08 07 06
20 09 08 07 06 05 04
PMP BJE

The sponsoring editor for this book was Harold B. Crawford, and the production supervisor was Suzanne Babeuf.

Library of Congress Cataloging-in-Publication Data

Taylor, Carson W.
Power system voltage stability / Carson W. Taylor.
p. cm.
EPRI Editors, Neal J. Balu and Dominic Maratukulam.
Includes bibliographical references and index.
ISBN 0-07-063184-0
1. Electric power system stability. I. Title.
TK1005.T29 1994
621.319-dc20 93-21455
CIP

When ordering this title, use ISBN 0-07-113708-4

Printed in Singapore

To my parents

Preface

Power transmission capability has traditionally been limited by either rotor angle (synchronous) stability or by thermal loading capabilities. The blackout problem has been associated with transient stability; fortunately this problem is now diminished by fast short circuit clearing, powerful excitation systems, and various special stability controls.

Voltage (load) stability, however, is now a major concern in planning and operating electric power systems. More and more electric utilities are facing voltage stability-imposed limits. Voltage instability and collapse have resulted in several major system failures (blackouts) such as the massive Tokyo blackout in July 1987.

Voltage stability will remain a challenge for the foreseeable future and, indeed, is likely to increase in importance. One reason is the need for more intensive use of available transmission facilities. The increased use of existing transmission is made possible, in part, by reactive power compensation—which is inherently less robust than "wire-in-the-air."

Over the last ten to fifteen years, and especially over about the last five years, utility engineers, consultants, and university researchers have intensely studied voltage stability. Hundreds of technical papers have resulted, along with many conferences, symposiums, and seminars. Utilities have developed practical analysis techniques, and are now planning and operating power systems to prevent voltage instability for credible disturbances. All relevant phenomena, including longer-term phenomena, can be demonstrated by time domain simulation.

While experts now have a good understanding of voltage phenomena, a comprehensive, practical explanation of voltage stability in book form is necessary. This is the first book on voltage stability.

Power System Voltage Stability is an outgrowth of many two–three day seminars which I began offering in 1988. As a full-time engineer of the Bonneville Power Administration (BPA), the book is influenced by my work on voltage stability problems in the Pacific Northwest and adjacent areas. It is also influenced by my participation in voltage stability work of the Western Systems Coordinating Council, North American Electric Reliability Council, IEEE, CIGRE, and EPRI.

Although voltage stability is fairly well understood, there are many facets to the problem, ranging from generator controls to transmission network reactive power compensation to distribution network design to load characteristics. The physical characteristics and mathematical models of a wide range of equipment are important.

Power System Voltage Stability emphasizes the physical or engineering aspects of voltage stability, providing a conceptual understanding of voltage stability. The simplest possible models are used for conceptual explanations. Practical methods for computer analysis are emphasized. We aim to develop good intuition relative to voltage problems, rather than to describe sophisticated mathematical analytical methods. The book is primarily for practicing engineers in power system planning and operation. However, the book should be useful to university students as a supplementary text. University researchers may find the book provides necessary background material on the voltage stability problem.

Many references are provided for those who wish to delve deeper into a fascinating subject. The references are not exhaustive, however, and generally represent recent publications which build on earlier work. In keeping with the intended audience, most of the references are quite readable by those without advanced mathematical training.

Outline of book. The book is divided into nine chapters and six appendices. Chapter 1 is introductory with emphasis on reactive power transmission. Chapter 2 introduces the subject of voltage stability, providing definitions and basic concepts. Voltage stability is separated into transient and longer-term phenomena.

Chapters 3–5 describe equipment characteristics for transmission systems, generation systems, and distribution/load systems. Modeling of equipment is emphasized.

Chapters 6 and 7 describe computer simulation examples for both small equivalent power systems and for a very large power system. Both static and dynamic simulation methods are used. Both transient and longer-term forms of voltage stability are studied using conventional and advanced computer programs.

Chapter 8 describes voltage stability associated with HVDC links. Here the reactive power demand of HVDC inverters is important.

Chapter 9 provides planning and operating guidelines, and potential solutions to voltage problems.

The appendices include description of computer methods for power flow and dynamic simulation, and description of voltage instability incidents.

Voltage stability is still a fresh subject and many advances in understanding, simulation software, and on-line security assessment software will be made in future years. In fact, the book was frequently updated until the submission deadline. It's likely that a revised edition will be called for. I invite comments on the book and suggestions for revised editions. Please write to me at 252 Northwest Seblar Court, Portland, Oregon 97210.

For those interested in desktop publishing, I used a Macintosh IIci computer and FrameMaker technical publishing software. I also used several other programs such as DeltaGraph and Canvas. The manuscript was submitted to McGraw-Hill on diskettes.

Acknowledgments. I am indebted to many seminar participants, BPA colleagues, and industry colleagues. To a large extent, the book is a synthesis of a large body of literature and practical knowledge. Through papers, correspondence, and discussions, Walter Lachs, Harrison Clark, Dr. Thierry Van Cutsem, Dr. Mrinal Pal, and others have provided many helpful insights. On an international level, I have been privileged to participate in the work of two CIGRÉ task forces investigating voltage stability. As part of EPRI software development projects, discussions with Dr. Prabha Kundur and colleagues at Ontario Hydro were very important.

Dr. Kundur and Mark Lauby reviewed the manuscript and provided many helpful suggestions. I, however, am solely responsible for the final version.

The Electric Power Research Institute sponsored the book. I deeply appreciate the support at EPRI from Dr. Neal Balu, Mark Lauby, and Dominic Maratukulam.

Although this was an off-hours project, I thank BPA engineering management for encouraging advances in power system engineering, for providing the opportunity to work on problems in the field of voltage stability, and for the privilege of participating in industry and professional society study of voltage stability. This work, however, is my own and does not necessarily reflect the views of the Bonneville Power Administration.

Finally, I thank my wife, Gudrun Taylor, for her encouragement, proofreading, and patience during many hours at the computer.

Carson W. Taylor
Portland, Oregon
December 1992

Foreword

Electric utilities have been forced in recent years to squeeze the maximum possible power through existing networks due to a variety of limitations in the construction of generation and transmission facilities.

Voltage stability is concerned with the ability of a power system to maintain acceptable voltages at all nodes in the system under normal and contingent conditions. A power system is said to have entered a state of voltage instability when a disturbance causes a progressive and uncontrollable decline in voltage.

Inadequate reactive power support from generators and transmission lines leads to voltage instability or voltage collapse, which have resulted in several major system failures in recent years. Hence, a thorough understanding of voltage stability phenomena and designing mitigation schemes to prevent voltage instability is of great value to utilities.

The author, Carson Taylor, is an internationally recognized expert on power system voltage stability. He not only has a thorough understanding of the fundamental concepts of voltage stability but also has demonstrated his skill in developing practical solutions to real life problems of voltage instability. Carson has taught many courses and written numerous technical papers on the subject of power system voltage stability.

It gives me great pleasure to write the Foreword for this timely book, which I am confident will be of great value to practicing engineers and students in the field of power engineering.

Dr. Neal J. Balu
Program Manager
Power System Planning and Operations Program
Electrical System Division

Contents

Appendices:

1

General Aspects of Electric Power Systems

Everything you know is easy.
Serbian saying

Power system voltage stability involves generation, transmission, and distribution. Voltage stability is closely associated with other aspects of power system steady-state and dynamic performance. Voltage control, reactive power compensation and management, rotor angle (synchronous) stability, protective relaying, and control center operations all influence voltage stability. Before introducing voltage stability in the next chapter, we review aspects of power system engineering important to power system planning and operating engineers.

1.1 Brief Survey of Power System Analysis and Operation

In this book, our overriding concern is power system security. We must avoid failures and blackouts of the bulk power delivery system. Economic system operation is of secondary importance during emergency conditions, but is important during normal conditions. In system design and operation we need a balance between economy and security.

Disturbances. A large interconnected power system is exposed to many disturbances which threaten security. Recent requirements for more intensive use of available generation and transmission have magnified the possible effects of these disturbances. For three-phase power systems, the disturbances can be divided into balanced and unbalanced disturbances. Unbalanced disturbances are normally caused by short circuits (faults) affecting only one or two of the phases; faults involving ground are the most common. Balanced disturbances result from transmission line and generation outages, and from load changes. Following any disturbance, electromechanical oscillations occur between generators.

Computer simulation programs. Large-scale computer simulation programs for studying power system steady-state and dynamic performance include short circuit programs, power flow programs, small-signal stability (eigenvalue) programs, transient stability programs, and longer-term dynamics programs.

Power flow programs are basic to power system analysis, planning, and operation. Similar network power flow computation techniques are used in other software for optimal power flow, dynamic simulation, on-line security assessment, and state estimation. Power flow programs normally represent the generation and transmission systems in the sinusoidal fundamental-frequency steady-state under balanced conditions. Loads are usually lumped at bulk power delivery substation busses. A solved case provides the voltage magnitudes and angles at each bus, and the real and reactive power flows. Appendix B describes the power flow problem.

Time domain transient stability programs are used to determine rotor angle synchronous stability performance—both the "first swing" and subsequent transient damping. The dynamic performance of induction motors and various controls can also be evaluated. Numerical integration is the computation method. Eigenvalue and related methods are also useful in evaluating electromechanical stability of linearized systems—damping of low-frequency oscillations and the effects of controls are often studied. These subjects are described in depth in the companion book *Power System Stability and Control* by Dr. Prabha Kundur.

Longer-term dynamics programs evaluate slower dynamics. These programs are discussed later chapters and in Appendix D.

Controls. Various power system controls—local and centralized—are important in voltage stability. The local controls, particularly at generating plants, are automatic and relatively high speed. Direct and indirect control of loads are critical for voltage stability.

Each company or control area has a central control or dispatch center where slower automatic and manual control commands are issued to power plants and substations. The primary centralized automatic control is Automatic Generation Control (AGC). Centralized voltage control usually has a "man-in-the-loop." Other than telephone communication, there is seldom central control at the synchronous interconnection level.

Large-scale systems. Electric power systems are the largest man-made dynamic systems on earth. Networks comprise thousands of nodes and the significant dynamics are equivalent to thousands of first-order nonlinear differential equations. At any instant in time, generation must match load. Generators thousands of kilometers apart connected by highly-loaded transmission circuits must operate in synchronism. This must be done reliably through the daily load cycle and for disturbance conditions.

Electric power systems are comprised of generation, transmission, and distribution/loads. These three subsystems must operate together as an overall system. We must understand each subsystem. Equally important, we must understand how they relate; this system engineering is shown by the intersection areas of Figure 1-1.

Voltage stability is only one aspect of power system engineering. But it is a very interesting one!

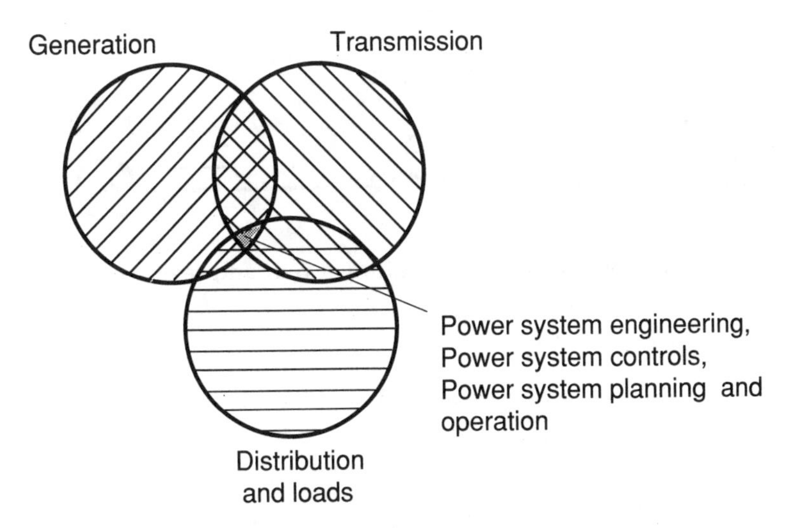

Fig. 1-1. Domains of power system engineering. With cogeneration and independent power producers, system engineering between generation and distribution is required.

1.2 Active Power Transmission using Elementary Models

In this section and the next, we review the basics of electric power transmission. To facilitate understanding, we use simple models. Once basic concepts are understood, we can represent as much detail as appropriate in computer simulation.

Active (real) and reactive power transmission depend on the voltage magnitude and angles at the sending and receiving ends. Figure 1-2 shows our model; synchronous machines are indicated at both ends. The sending- and receiving-end voltages are assumed fixed and can be interpreted as points in large systems where voltages are stiff or secure. The sending and receiving ends are connected by an equivalent reactance.

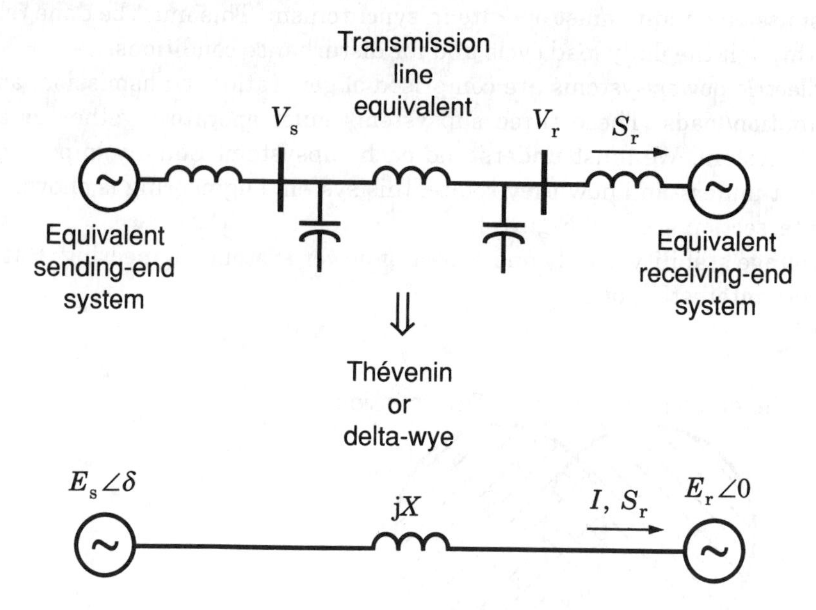

Fig. 1-2. Elementary model for calculation of real and reactive power transmission.

The relations can be easily calculated:

$$\bar{S}_r = P_r + jQ_r = \bar{E}_r I^*$$

$$= E_r \left[\frac{E_s \cos\delta + jE_s \sin\delta - E_r}{jX} \right]^*$$

$$= \frac{E_s E_r}{X} \sin\delta + j \left[\frac{E_s E_r \cos\delta - E_r^2}{X} \right]$$

$$P_r = \frac{E_s E_r}{X} \sin\delta = P_{max} \sin\delta \tag{1.1}$$

$$Q_r = \frac{E_s E_r \cos\delta - E_r^2}{X} \tag{1.2}$$

Similarly, for the sending end:

$$P_s = \frac{E_s E_r}{X} \sin\delta = P_{max} \sin\delta \tag{1.3}$$

$$Q_r = \frac{E_s^2 - E_s E_r \cos\delta}{X} \tag{1.4}$$

The familiar equations for P_s and P_r are equal because we have a lossless system; maximum power transfer is at a power or load angle δ equal to 90°. Figure 1-3 shows the plot of Equation 1.1 or 1.3, termed the power-angle curve. The 90° maximum power angle is nominal—maximum power occurs at a different angle if we included transmission losses or resistive shunt loads. In the next chapter, we will describe the case where the receiving end is an impedance load.

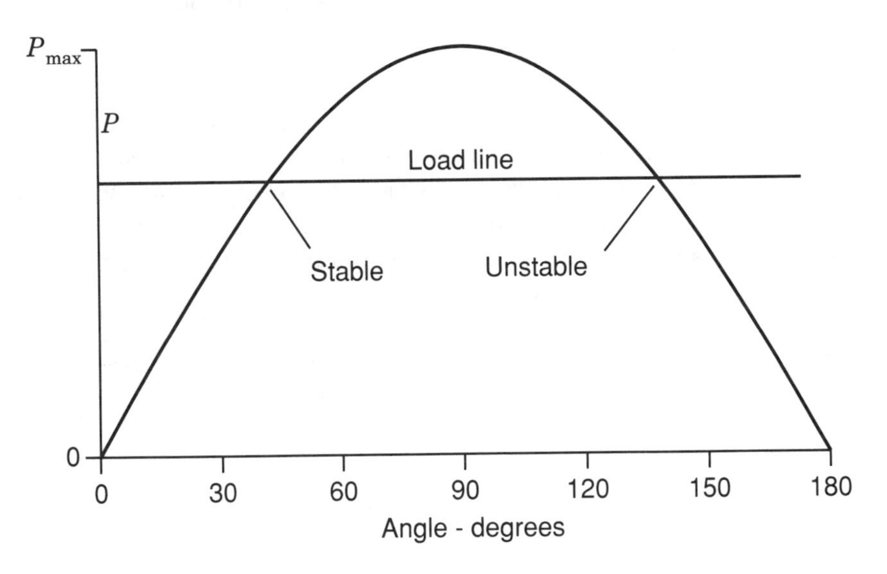

Fig. 1-3. Power angle curve for elementary model.

Also shown on Figure 1-3 is a load line representing constant sending end mechanical (turbine) power. The stable intersection of the mechanical power and the electrical power-angle curve is on the left at an angle less than 90°. The right side intersection is unstable. To understand this, let the receiving-end system be very large, an infinite bus with fixed angle and speed. A small increase in mechanical power at the sending-end generator (by opening the steam valves or water gates) accelerates the generator, thereby increasing the angle. The increased angle results in less electrical power which further accelerates the generator and further increases the angle. On the left-side intersection, however, the increased angle increases the electrical power to match the increased mechanical power.

For typical power transfers and power angles, say up to 30°, we can linearize Equations 1.1 and 1.3 by the relation $\sin\delta \cong \delta$ with δ in radians. (For example, thirty degrees is 0.5236 radians and sin 30° = 0.5; note the nearly linear relationship for small angles on Figure 1-3.) We then write:

$$P \cong P_{max}\,\delta$$

and state: *Real or active power transfer depends mainly on the power angle.*

To insure steady-state rotor angle (synchronous) stability, angles across a transmission system are usually kept below about 44° [1].

1.3 Reactive Power Transmission using Elementary Models

In this book, we are especially interested in the transmission of reactive power. First, we can return to the power-angle curve and note that the reactive requirements of the sending and receiving ends are excessive at high angles and correspondingly high real power transfers. Making the assumption that $E_s = E_r$, Figure 1-4 shows a plot for $Q_s = -Q_r$. At the 90° steady-state stability limit, the reactive power that must be generated at the two sources is equal to P_{max}. (Referring to Figure 1-2, part of the reactive power is provided by transmission line capacitance).

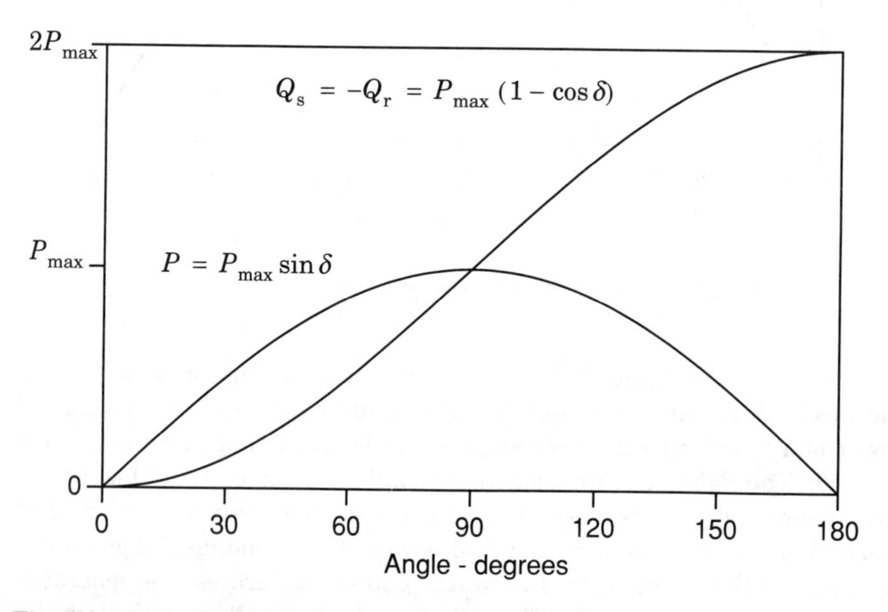

Fig. 1-4. Power-angle curve for elementary model. Also shown is the curve for reactive power with $E_s = E_r$.

Usually we are interested in variable voltage magnitudes. Particularly, we are interested in the reactive power that can be transmitted across a transmission line, a transmission line equivalent, or a transformer as the receiving- or load-end voltage sags during a voltage emergency or collapse. Referring to Figure 1-2, we now consider reactive power flow over the transmission line alone and rewrite Equations 1.2 and 1.4 in terms of $V_s\angle\theta$ and $V_r\angle\theta$. Also, X now represents the transmission line reactance alone.

$$Q_r = \frac{V_s V_r \cos\theta - V_r^2}{X} \tag{1.5}$$

$$Q_s = \frac{V_s^2 - V_s V_r \cos\theta}{X} \tag{1.6}$$

We can write approximate formulas for small angles by using $\cos\theta \cong 1$:

$$Q_s = \frac{V_r (V_s - V_r)}{X} \tag{1.7}$$

$$Q_s = \frac{V_s (V_s - V_r)}{X} \tag{1.8}$$

From Equations 1.7–1.8, we state: *Reactive power transmission depends mainly on voltage magnitudes and flows from the highest voltage to the lowest voltage*. Also:

- *P and δ are closely coupled*, and
- *Q and V are closely coupled.*

These physical relationships are taken advantage of in computer algorithms, notably the fast decoupled power flow (Appendix B).

Next, we examine how these relationships break down during high stress, i.e., high power transfers and angles. This is important since voltage stability problems normally occur during highly stressed conditions (usually following outages).

Example 1-1. First let's simply calculate Q_s and Q_r using Equations 1.5 and 1.6 for an angle of 30°. Let V_s be 1 per unit and let V_r be 0.9 per unit. We have a substantial voltage gradient of 10% between the two ends and we might expect large reactive power transfer. We calculate:

$$\cos 30^\circ = 0.866$$

$$Q_s = \frac{1^2 - 1 \cdot 0.9 \cdot 0.866}{X} = \frac{0.22}{X}$$

$$Q_r = \frac{1 \cdot 0.9 \cdot 0.866 - 0.9^2}{X} = -\frac{0.03}{X}$$

We have a problem! Although lots of reactive power is going into the line, nothing is coming out. The negative value means, in fact, that the line is demanding reactive power of $0.03/X$ pu from the receiving end. The transmission line has become a *drain* on the transmission system. The transmission line reactive loss is the sum of the reactive powers going into the line or $0.25/X$ pu.

Power circle diagrams. P, Q circle diagrams are a more general and precise way of understanding power transmission limitations. Circle diagrams were widely used prior to digital computer power flow programs, and are described in several textbooks on power system analysis and in the Westinghouse T&D book [2]. The next example demonstrates this method.

Example 1-2. A 500-kV transmission line is 161 km (100 miles) long. We include the line shunt capacitance using a pi transmission line model. For simplicity, and with little error, we can let the line be lossless. Corresponding to an actual Bonneville Power Administration line with three subconductor per phase, the total series reactance is 51.6 ohms and the total shunt susceptance is 809 microsiemens (micromhos). In per unit on a 500-kV and 1000 MVA base, the reactance and susceptance parameters are 0.2064 per unit and 0.2023 per unit, respectively. (One thousand megawatts is approximately the 500-kV line natural or surge impedance loading and is a convenient power base; the impedance base is 250 ohms which is approximately the surge impedance.) Using ABCD generalized circuit constants [2] for the pi model, the parameters are:

$$\bar{A} = \bar{D} = 1 + \frac{\bar{Z}\bar{Y}}{2} = 0.9791 \text{ pu}$$

$$\bar{B} = \bar{Z} = \mathrm{j}0.2064$$

$$\bar{C} = \bar{Y} + \frac{\bar{Z}\bar{Y}^2}{4} = \mathrm{j}0.2002$$

We consider two cases: Case 1 has $V_s = 1$ per unit and $V_r = 0.95$ per unit; Case 2 has $V_s = 1$ per unit and $V_r = 0.9$ per unit. Because the line is lossless, the center of all sending- and receiving-end circles are on the vertical reactive power axis. The two cases result in two sending end and two receiving circles. The circle centers and radii (same for both sending and receiving ends) are:

$$Center_{send} = -\frac{\overline{D}}{\overline{\overline{B}}}V_s^2 = j4.7437 \text{ pu}$$

$$Center_{rec} = \frac{\overline{A}}{\overline{\overline{B}}}V_r^2 = -j4.2812 \text{ for } V_r = 0.95 \text{ pu}$$

$$Center_{rec} = \frac{\overline{A}}{\overline{\overline{B}}}V_r^2 = -j3.8451 \text{ pu for } V_r = 0.9 \text{ pu}$$

$$Radius = \frac{V_s V_r}{\overline{\overline{B}}} = -4.6028 \text{ pu for } V_r = 0.95 \text{ pu}$$

$$Radius = \frac{V_s V_r}{\overline{\overline{B}}} = -4.3605 \text{ pu for } V_r = 0.9 \text{ pu}$$

Figure 1-5 shows the power circle curves. The solid circles are for Case 1 and the dashed circles are for Case 2. For any specified real power transfer, we can draw a vertical line. The intersection of the vertical line and a circle gives reactive power. Also, for a specified real power, the angle between the vertical axis and a line drawn from a circle center to the point on the circle is the power angle θ.

Concentrating on the receiving-end circles, we clearly see when reactive power becomes negative—and the transmission line becomes a drain on the receiving-end system. For $V_r = 0.95$ per unit, the power value is about 1700 MW. For $V_r = 0.9$ per unit, the power value is about 2250 MW. We also note the high reactive power requirements from the sending- and receiving-end systems at very high real power transfers. The corresponding angles can be determined graphically or analytically (Equation 1.1).

At high loadings the curves become steep, meaning that more than one megavar is required for each additional megawatt transmitted.

The same method can be used for series/parallel combinations of transmission lines. Generalized circuit constants (ABCD constants) facilitate converting the lines to a single pi equivalent.

1.4 Difficulties with Reactive Power Transmission

The last section hinted at one difficulty with reactive power transmission: Reactive power cannot be transmitted across large power angles even with substantial voltage magnitude gradients. High angles are due to long lines and high real power transfers. Requirements to maintain voltage magnitude profiles with voltages of approximately 1 per unit ± 5% contribute to

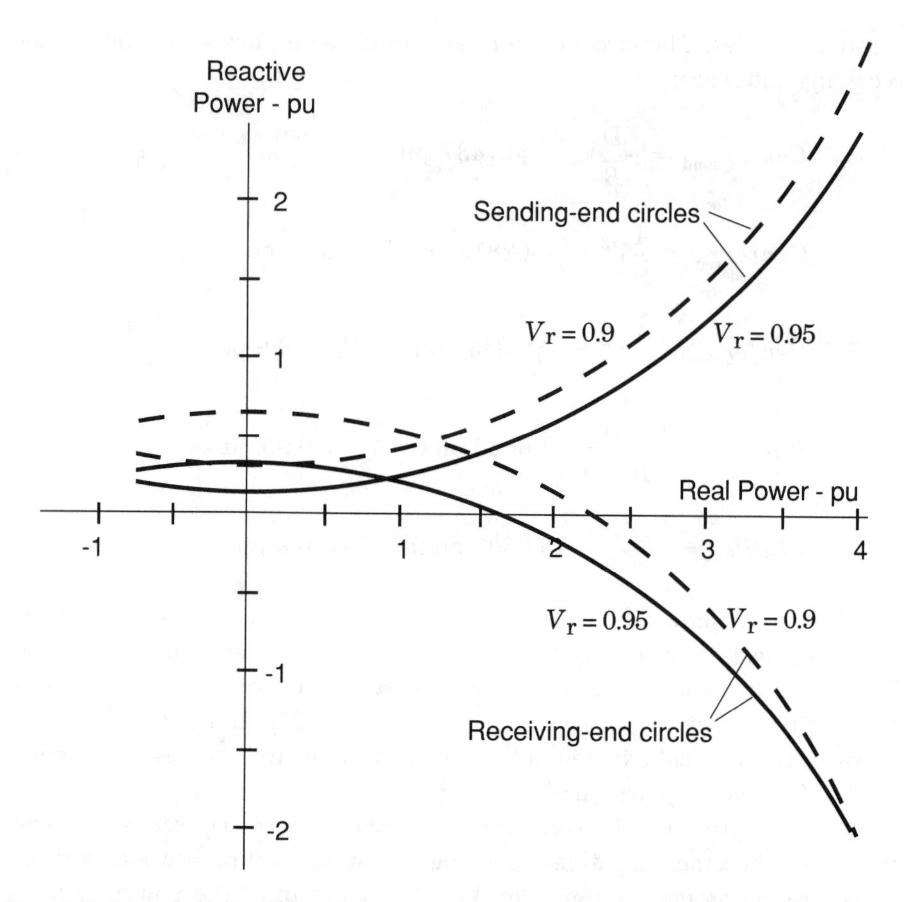

Fig. 1-5. Power circle diagram for a 161 km, 500-kV transmission line with one per unit sending-end voltage and depressed receiving-end voltages. One per unit power is 1000 MVA.

the difficulty. Contrasted with real power transfers, reactive power simply cannot be transmitted long distances.

There are other reasons to minimize transfer of reactive power. Minimizing real and reactive losses is a second reason. Real losses should be minimized for economic reasons; reactive losses should be minimized to reduce investment in reactive power devices such as shunt capacitors.

The losses across the series impedance of a transmission line are I^2R and I^2X. For I^2, we can write:

$$I^2 = \bar{I} \cdot \bar{I}^* = \left[\frac{P - jQ}{\bar{V}^*}\right]\left[\frac{P + jQ}{\bar{V}}\right] = \frac{P^2 + Q^2}{V^2}$$

and

$$P_{loss} = I^2 R = \frac{P^2 + Q^2}{V^2} R \quad (1.10)$$

$$Q_{loss} = I^2 X = \frac{P^2 + Q^2}{V^2} X \quad (1.11).$$

To minimize losses, we must minimize reactive power transfer. We should also keep voltages high. Keeping voltages high to minimize reactive losses helps maintain voltage stability.

Minimizing temporary overvoltage due to "load rejection" is a third reason. The most onerous case is opening the receiving-end circuit breakers with the transmission line still energized from the sending end.

Figure 1-6 shows an equivalent system and an even simpler thévenin circuit. Also shown is the resulting phasor diagram. Prior to the breaker opening, the thévenin voltage is:

$$E_{th} \angle \delta = V \angle 0 + jX\bar{I} = V + jX\frac{P_r - jQ_r}{V} = V + \frac{XQ_r}{V} + j\frac{XP_r}{V} \quad (1.12).$$

From the equation and phasor diagram, we note that the voltage rise term in phase with V depends on Q. This term mainly determines E_{th} (the thévenin voltage magnitude). The angle, δ, depends mainly on the quadrature term involving P.

What happens when the breaker is opened? What does the voltage at the open end of the line become? Clearly, the current goes to zero and the voltage becomes E_{th}. Thus the temporary overvoltage is largely determined by the reactive power transfer. Two examples will show this.

Example 1-3. A 100 km, 500-kV line has a series reactance of $X = 0.35\ \Omega$/km. The power transmitted is 1000 MW or 1 per unit. The impedance base is 250 ohms and the transmission line reactance is 0.14 per unit. At the source end, the short circuit capacity* is 5000 MW or 5 per unit; the corresponding source reactance is 1/5 or 0.2 per unit. The total reactance is 0.34 per unit. The line is relatively short but the source is relatively weak. Let V = 1 per unit. Consider two cases: unity power factor load and 0.85 power factor load.

Case 1: $\cos \phi = 1$, $Q_r = 0$,

---

*Short circuit capacity is discussed in the next section.

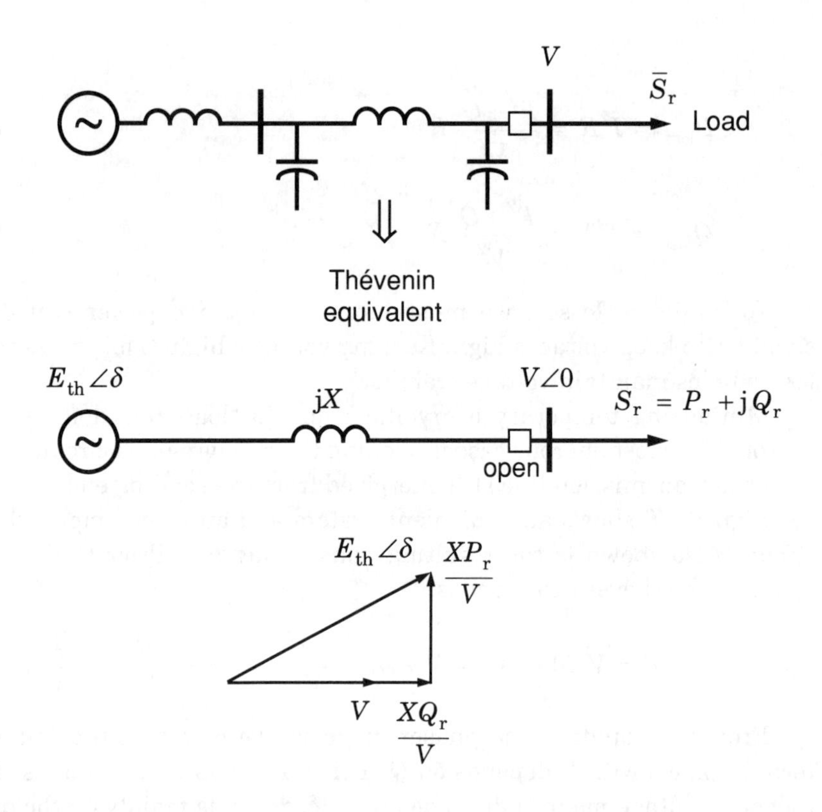

Fig. 1-6. Equivalent system for load rejection calculation.

$$E_{th}\angle\delta = V + j\frac{XP_r}{V} = 1 + j\frac{0.34 \cdot 1}{1} = 1.056\angle 18.8°$$

Case 2: cos ϕ = 0.85, $Q_r = P_r \tan\phi = 0.62$ pu,

$$E_{th}\angle\delta = 1 + (0.34 \cdot 0.62) + j\,(0.34 \cdot 1) = 1.211 + j0.341$$

$$= 1.258\angle 15.7°$$

The delivery of 0.62 pu reactive power has increased the fundamental frequency load rejection overvoltage from 1.056 per unit to 1.258 per unit. The next example is more dramatic.

Example 1-4. This example involves a high voltage direct current (HVDC) transmission link connected to a weak power system. The thévenin reactance is 0.625, corresponding to the inverse of the "effective short circuit ratio" (ESCR) which is 1.6. Several existing HVDC links have such a high network reactance. HVDC converters consume reactive power of 50–60% of the dc power. Let the dc power be 1 per unit and the reactive consumption

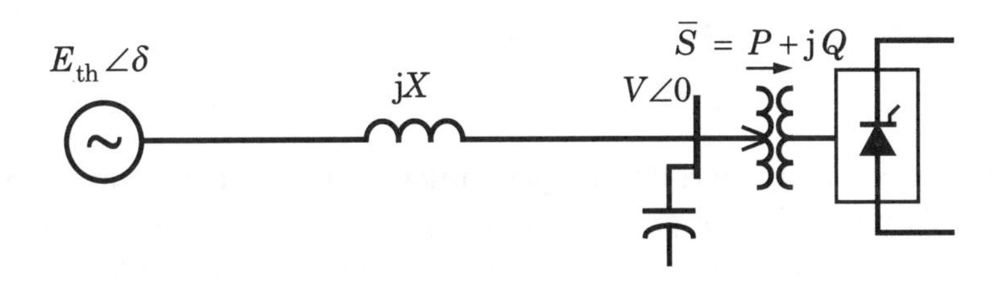

Fig. 1-7. Example 1-3, HVDC link fed from high reactance network.

be 0.6 per unit; also, let the converter commutating bus voltage be 1 per unit. See Figure 1-7. Using Equation 1.12, the thévenin voltage is:

$$E_{th}\angle\delta = 1 + (0.625 \cdot 0.6) + j\,(0.625 \cdot 1) = 1.51\angle 24°$$

For dc shutdown or blocking, the ac side voltage will jump to an intolerable level of 1.51 per unit! Measures must be taken to prevent such a high load rejection voltage. (One measure would be to produce the reactive power locally with a synchronous condenser—this increases the effective short circuit capacity and correspondingly reduces the thévenin reactance.) Note that if the dc converter is an inverter, the same problem occurs; only the signs of the dc power and of the thévenin angle are changed. Rectifiers and inverters consume similar amounts of reactive power.

Whenever possible, reactive power should be generated close to the point of consumption. We can list several reasons to minimize reactive power transfer:

1. It is inefficient during high real power transfer and requires substantial voltage magnitude gradients.
2. It causes high real and reactive power losses.
3. It can lead to damaging temporary overvoltages following load rejections.
4. It requires larger equipment sizes for transformers and cables.

1.5 Short Circuit Capacity, Short Circuit Ratio, and Voltage Regulation

We now discuss some terms used in the previous section. These are useful in describing the voltage (as opposed to mechanical or inertial) strength of a network. The concepts are useful for simple calculations prior to computer studies.

Short circuit capacity. The short circuit capacity or power of a network is the product of three-phase fault current and rated voltage. In physical units, with short circuit current in kiloamperes and phase-to-phase voltage in kilovolts, the short circuit capacity is:

$$S_{sc} = \sqrt{3} \times V \times I \ \text{MVA}$$

We, however, will generally use per unit quantities. The short circuit capacity is then simply the product of voltage (usually one per unit) and fault current. The fault current is usually considered to be rated (one per unit) voltage divided by the impedance or reactance to the fault location. With one per unit voltages, the short circuit capacity is then the system admittance (or susceptance), or the inverse of the system thévenin impedance (or reactance).

The short circuit capacity and thévenin impedance can be computed with a short circuit program. A transient stability program can also be used by applying a three-phase fault and noting the initial current flow to the fault point.

The short circuit capacity measures the system voltage strength. A high capacity (and corresponding low impedance) means the network is strong or stiff. Switching on a load, or a shunt capacitor or reactor, will not change the voltage magnitude very much. A low short circuit capacity means the network is weak.

Short circuit ratio (SCR). We sometimes wish to compare the size of equipment to the strength of the power system. The equipment could be a load (such as a large motor), an HVDC converter, or a static var compensator. A simple comparison is to divide the system strength by the device size. Comparing a 1000 MW HVDC converter to 5000 MVA short circuit capacity power system results in a short circuit ratio of 5000/1000 or 5. A high short circuit ratio means good performance. A low short circuit ratio means trouble: for example, a large motor connected to a weak point on the network may stall or have difficulty reaccelerating following faults. Motor starting will cause system voltage dips.

A related term, used especially with HVDC, is effective short circuit ratio (ESCR). The basic SCR accounts for only the network strength while ESCR accounts for shunt reactive equipment at the device location. A synchronous condenser clearly increases the fault current and therefore the effective short circuit capacity. On the other hand, shunt capacitors and harmonic filters (which are capacitive at fundamental frequency) reduce the ESCR.

Short circuit capacity related measures do not account for the fast-acting controls of static var compensators, generator voltage regulators, and HVDC converters. Methods which include control effects are described in Chapter 8.

Voltage regulation. Several widely used approximate formulas involve system short circuit capacity [3, Chapter 1]. The formulas provide the voltage deviation for switching shunt reactive equipment. They are:

$$\Delta V \cong \frac{\Delta Q}{S_{sc}} \qquad (1.13).$$

and

$$V \cong E\left[1 - \frac{Q}{S_{sc}}\right] \qquad (1.14).$$

Example 1-5. A 200-MVAr capacitor is switched at a bus with 10,000 MVA short circuit capacity. The expected voltage change is 200/10,000 or 2%. This approximate result can be checked by computer power flow or stability simulation.

Example 1-6. A bus experiences ±3% voltage fluctuations. The short circuit capacity is 5000 MVA. We wish to size a static var compensator (SVC) to smooth the voltage fluctuations. Using Equation 1.13, we can write: $\Delta Q \cong S_{sc} \Delta V$. The approximate SVC size is then ±150 MVAr.

Equation 1.14 approximates an even simpler equation expressing the voltage drop from the source to load point. That is: $V = E - jXI$. Figure 1-8 is a plot of Equation 1.14 and shows the system voltage/reactive character-

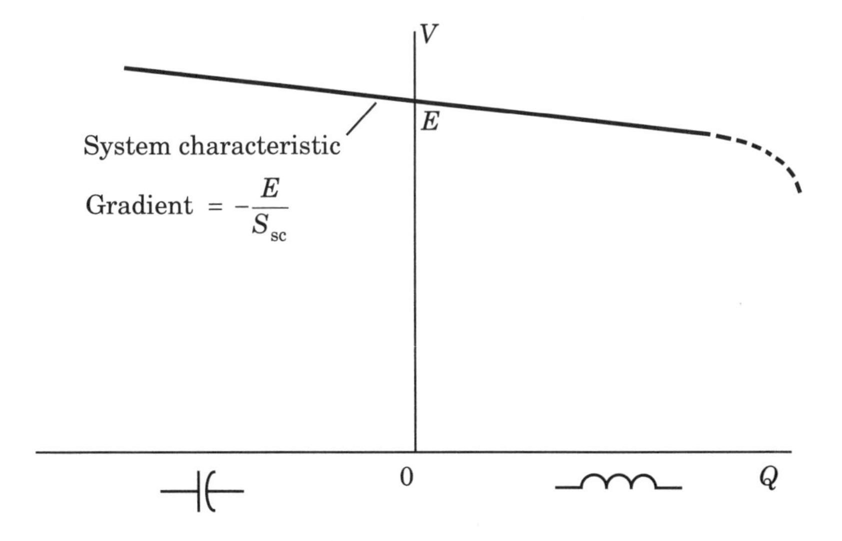

Fig. 1-8. System approximate voltage/reactive power characteristic.

istic or load line. The slope of the load line is related to the system stiffness— a nearly flat slope means a strong system. The voltage/reactive characteristics of shunt reactive devices (capacitor, reactor, or static var compensator) can be superimposed on the system characteristic.

In the next chapter, we introduce *V–Q* curves produced by computer power flow simulation. The system characteristic will not be linear at high inductive loading. The dashed portion of Figure 1-8 gives a hint of voltage problems.

References

1. R. D. Dunlop, R. Gutman, and P. P. Marchenko, "Analytical Development of Loadability Characteristics for EHV and UHV Transmission Lines," *IEEE Transactions on Power Apparatus and Systems*, Vol. PAS-98, No. 2, pp. 606–617, March/April 1979.
2. Westinghouse Electric Corporation, *Electrical Transmission and Distribution Reference Book*, East Pittsburgh, Pennsylvania, 1964.
3. T. J. E. Miller, editor, *Reactive Power Control in Electric Systems*, John Wiley & Sons, New York, 1982.

2

What is Voltage Stability?

Make everything as simple as possible, but not more so.
A. Einstein

We now introduce *voltage stability*, the subject of this book. First, we present some definitions. Then, we describe voltage instability mechanisms and the relation with rotor angle stability. Next, we discuss the reasons for voltage stability problems in mature power systems. We finish by introducing static voltage stability analysis using P–V and V–Q curves.

2.1 Voltage Stability, Voltage Collapse, and Voltage Security

Voltage stability covers a wide range of phenomena. Because of this, voltage stability means different things to different engineers. It's a fast phenomenon for engineers involved with induction motors, air conditioning loads, or HVDC links. It's a slow phenomenon (involving, for example, mechanical tap changing) for other engineers. Engineers and researchers have discussed appropriate analysis methods, with debate on whether voltage stability is a static or dynamic phenomenon.

Voltage instability and voltage collapse are used somewhat interchangeably by most engineers.

Voltage stability or voltage collapse has often been viewed as a steady-state "viability" problem suitable for static (power flow) analysis. The ability to transfer reactive power from production sources to consumption sinks during steady operating conditions *is* a major aspect of voltage stability. A 1987 CIGRÉ report [1] recommends analysis methods and power system planning approaches based on static models.

The network maximum power transfer limit is not necessarily the voltage stability limit.

Voltage instability or collapse is a dynamic process. The word "stability" implies a dynamic system. A power system *is* a dynamic system. We will see that, in contrast to rotor angle (synchronous) stability, the dynamics mainly involves the loads and the means for voltage control. Voltage stability has been called *load stability* [2].

Definitions. Voltage stability is a subset of overall power system stability. We adopt definitions developed by CIGRÉ [3]. The definitions are based on reference 4 and are in the spirit of references 5–7. The stability definitions are analogous to stability definitions for other dynamic system. Our definitions are:

> A power system at a given operating state is *small-disturbance voltage stable* if, following any small disturbance, voltages near loads are identical or close to the pre-disturbance values. (Small-disturbance voltage stability corresponds to a related linearized dynamic model with eigenvalues having negative real parts. For analysis, discontinuous models for tap changers may have to be replaced with equivalent continuous models.)
>
> A power system at a given operating state and subject to a given disturbance is *voltage stable* if voltages near loads approach post-disturbance equilibrium values. The disturbed state is within the region of attraction of the stable post-disturbance equilibrium.*
>
> A power system at a given operating state and subject to a given disturbance undergoes *voltage collapse* if post-disturbance equilibrium voltages are below acceptable limits. Voltage collapse may be total (blackout) or partial.

Voltage instability is the absence of voltage stability, and results in progressive voltage decrease (or increase). Destabilizing controls reaching limits, or other control actions (e.g., load disconnection), however, may establish global stability.

Voltage stability normally involves large disturbances (including rapid increases in load or power transfer). Furthermore the instability is almost always an aperiodic decrease in voltage. Oscillatory voltage instability may be possible [8], but control instabilities are excluded. Control instabilities could occur, for example, because of too high a gain on a static var compensator or too small a deadband in a voltage relay controlling a shunt capacitor bank. Overvoltage phenomena and instability such as self-excitation of rotating machines are outside the scope of the definitions. Overvoltages are

---

*Equilibrium points and regions of attraction are described in the next section.

normally more of an equipment problem than a power system stability problem.

The term *voltage security* is used. This means the ability of a system, not only to operate stably, but also to remain stable following credible contingencies or load increases [9]. It often means the existence of considerable margin from an operating point to the voltage instability point (or to the maximum power transfer point) following credible contingencies.

Although voltage stability involves dynamics, power flow based static analysis methods are often useful for rapid, approximate analysis.

2.2 Time Frames for Voltage Instability, Mechanisms

Voltage instability and collapse dynamics span a range in time from a fraction of a second to tens of minutes. Time response charts have been used to describe dynamic phenomena [10,11]. Figure 2-1 shows that many power system components and controls play a role in voltage stability. Only some, however, will significantly participate in a particular incident or scenario. The system characteristics and the disturbance will determine which phenomena are important.

Figure 2-1 also shows a classification of voltage stability into transient and longer-term time frames. There is almost always a clear separation

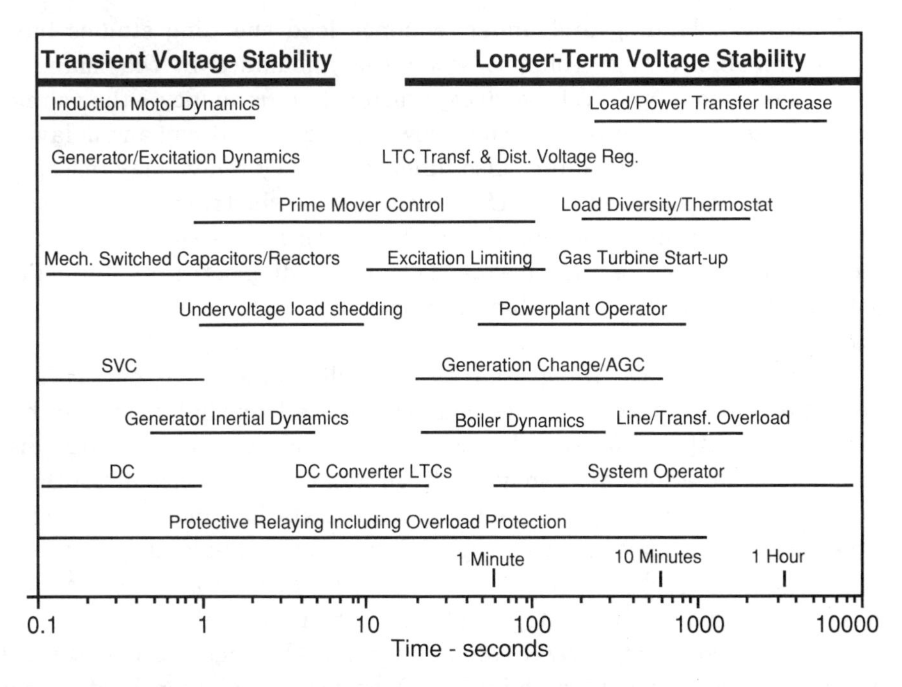

Fig. 2-1. Voltage stability phenomena and time responses.

between the two time frames. Actual incidents experienced by utilities are grouped by time frame in Appendix F.

Mechanisms—scenarios: We now describe the two classifications of voltage instability. Only the basic ideas are described in this introduction. We describe three scenarios.

1. *Scenario 1: transient voltage stability.* The time frame is zero to about ten seconds—which is also the time frame of transient rotor angle stability. The distinction between voltage instability and rotor angle instability isn't always clear, and aspects of both phenomena may exist. Does voltage collapse cause loss of synchronism, or does loss of synchronism cause voltage collapse? Voltage collapse is caused by unfavorable fast-acting load components such as induction motors and dc converters.

For severe voltage dips (such as during slowly-cleared short circuits), the reactive power demand of induction motors increases, contributing to voltage collapse unless protection or ac contactors trip the motors. (This has also been termed induction motor instability [12].) Following faults, motors have difficulty reaccelerating. Stall-prone motors can cause other nearby motors to stall. In simulation studies, motors must be represented as dynamic devices. The characteristic of shunt capacitor bank compensation (reactive power proportional to the voltage squared) adds to the problems.

Electrical islanding and underfrequency load shedding studies have shown probable voltage collapse when the imbalance in the island is greater than about 50%. Voltage decays faster than frequency—the voltage decay affects voltage-sensitive loads, slowing frequency decay and delaying underfrequency load shedding. Also, underfrequency relays may not operate because of the low voltages. Undervoltage load shedding may be necessary. For an incident in Florida, Figure 2-2 shows voltage collapsing before frequency decays to the underfrequency load shedding setpoints. Induction motor loads, including power plant auxiliary motors, were important in the incident.

In recent years, the integration of high voltage direct current (HVDC) links into voltage-weak power systems has caused transient voltage stability problems [14]. As an example, for stressed conditions and for large disturbances, simulations have shown voltage collapse tendencies in Southern California, aggravated by the two large inverter stations near Los Angeles. Sometimes (at the expense of synchronizing power) it's necessary to reduce dc power (and thereby converter reactive power demand) to support voltages. Chapter 8 describes voltage stability with HVDC links.

2. *Scenario 2: longer-term voltage stability.* The time frame is several minutes, typically two–three minutes. Operator intervention is often not

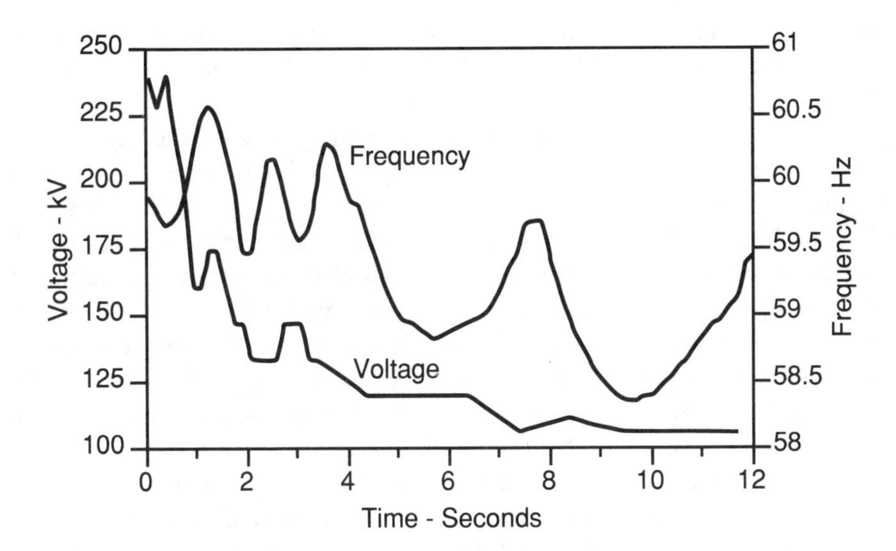

Fig. 2-2. Voltage and frequency for South Florida blackout on May 17, 1985 [13]. (See Appendix F for further description.)

possible. The terms "mid-term" stability, and "post-transient" or "post-disturbance" stability have been used.

This scenario involves high loads, high power imports from remote generation, and a sudden large disturbance. The system is transiently stable because of the voltage sensitivity of loads. The disturbance (loss of large generators in a load area or loss of major transmission lines) causes high reactive power losses and voltage sags in load areas. Tap changers on bulk power delivery LTC* transformers and distribution voltage regulators sense the low voltages and act to restore distribution voltages—thereby restoring load power levels.

The load restoration causes further sags of transmission voltages. Nearby generators are overexcited and overloaded, but overexcitation limiters (or power plant operators) return field currents to rated values as the time-overload capability (one to two minutes) expires. Generators farther away must then provide the reactive power. As described in Chapter 1, this is inefficient and ineffective. The generation and transmission system can no longer support the loads and the reactive losses, and rapid voltage decay ensues. Partial or complete voltage collapse follows. The final stages

*The terms LTC (Load Tap Changing), ULTC (Under-load Tap Changing), and OLTC (On-Load Tap Changing) are widely used. We use "LTC" since it is the term specified in *IEEE Standard Dictionary of Electrical and Electronic Terms*, ANSI/IEEE Std 100-1988.

may involve induction motor stalling and protective relay operations. Depending on the type of loads (including means for disconnection at low voltage) the collapse may be partial or total.

3. *Scenario 3: longer-term voltage instability.* The instability evolves over a still longer time period and is driven by a very large load buildup (morning or afternoon pickup), or a large rapid power transfer increase. The load buildup, measured in megawatts/minute, may be quite rapid. Operator actions, such as timely application of reactive power equipment or load shedding, may be necessary to prevent instability. Factors such as the time-overload limit of transmission lines (tens of minutes) and loss of load diversity due to low voltage (due to constant energy, thermostatically controlled loads) may be important. The final stages of instability involve actions of faster equipment as described for scenarios 1 and 2.

There are many interactions among the various equipment (Figure 2-1) and time frames. For example, tap changer regulation of voltages will prevent loss of diversity by thermostatic regulation of constant energy loads. For another example, overexcitation limiter operation prevents normal generator voltage regulation.

Mechanisms—load dynamics, equilibrium points, and region of attraction.*[*] Voltage stability has been called load stability. The "load" is the load seen at transmission system high voltage busses and includes the effects of subtransmission and distribution systems. The restoration of loads that have been temporary reduced because of low voltage is a key aspect of voltage stability.

Active (real) load is restored by three mechanisms:

1. Induction motors respond rapidly to match their mechanical load within a few seconds following sudden changes in voltage. Immediately following a sudden change in the source system, induction motors acts as impedance loads; this is apparent from the well-known equivalent circuit, considering that slip cannot change instantaneously because of motor inertia. For slow voltage decay, fast-responding motors track the slow dynamics of other equipment, acting as constant active power loads.
2. Automatic tap changing on bulk power delivery transformers and distribution voltage regulators operates over tens of seconds to several minutes to restore load-side voltage, and thus

---

*If this subsection is difficult, skim it and return after reading the rest of this chapter and Chapter 4. References 5, 6, and 15 provide expanded explanation.

voltage sensitive loads. Reactive power load and reactive power output of shunt compensation are also restored.

3. Constant energy resistive loads are restored by thermostatic or manual control. For aggregated loads this results in a loss of load diversity over a period of time following a voltage reduction.

As a first approximation, the dynamics of all three load restoration mechanisms are first order and can be modeled using a single time constant. (For tap changing, involving deadbands, voltage and timer relays, and discrete taps, the approximation is quite crude.) Although the time constants are different, the equations are similar in form and the load restoration mechanisms can be unified for conceptual analysis [6,7]. The three types of loads may be present at one bulk power delivery load bus as shown on Figure 2-3.

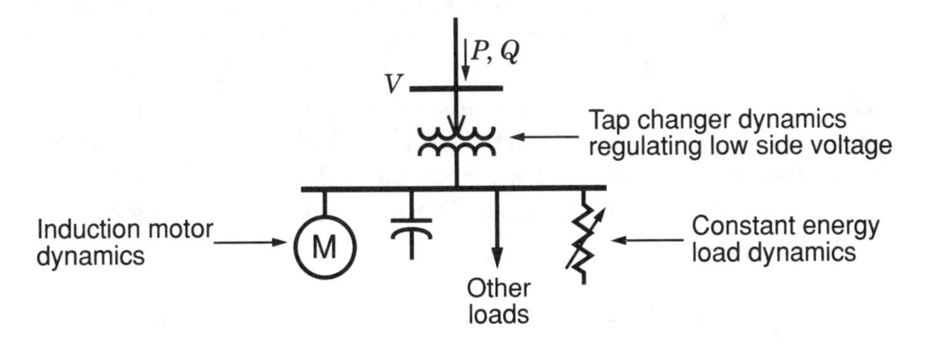

Fig. 2-3. Three mechanisms for restoration of voltage sensitive loads [7].

The state variables for first order models of the three load restoration mechanisms can be taken, respectively, as motor slip (s), tap changer turns ratio (n), and load conductance (G). As each state variable increases from zero, load power increases, reaches a maximum, and then decreases (Figure 2-4). As the state variable increases, voltage monotonically decreases.

For induction motors, Figure 2-4 is similar to familiar torque-slip curves. The applicable first order differential equation is:

$$2H\omega\frac{ds}{dt} = P_0 - P_e \qquad (2.1)$$

where P_0 is the initial mechanical power which, for simplicity, is assumed constant.

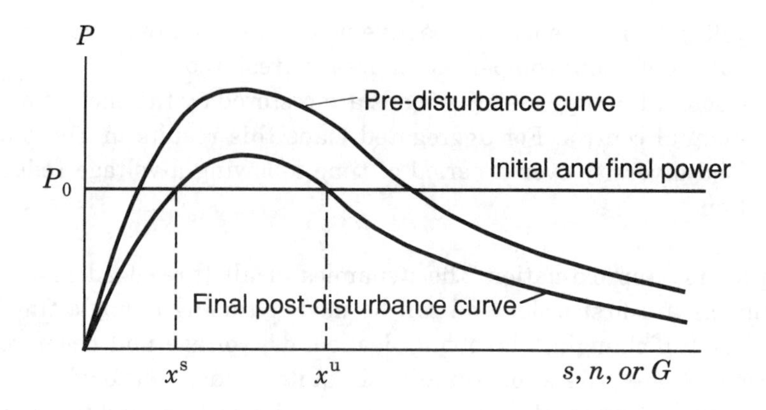

Fig. 2-4. Power versus state variable of load dynamics. Adapted from Van Cutsem [7].

For stability following a large disturbance, the slip at the moment of the final source system configuration (say, after fault clearing) must be within the "region of attraction" of the stable equilibrium point, x^s. The region of attraction extends to the unstable equilibrium point, x^u. Refer to Equation 2.1. For the region between x^s and x^u, this is because P_e is greater than P_0 and the motor will accelerate (slip will decrease) to x^s.

For a large number of thermostatically-controlled heating loads, an applicable equation is:

$$T\frac{dG}{dt} = P_0 - V_L^2 G \tag{2.2}$$

The situation is similar to the induction motor case. For stability following a large disturbance, the conductance at the moment of the final source system configuration (say, after circuit restoration and shunt capacitor insertion) must be within the "region of attraction" of the stable equilibrium point, x^s. The region of attraction extends to the unstable equilibrium point, x^u. Refer to Equation 2.2. For the region between x^s and x^u, this is because $V_L^2 G$ is greater than P_0 and the thermostats will reduce conductance (and increase voltage) until point x^s is reached.

The situation with tap changers is again similar. In fact, the formulation could be changed so that the Equation 2.2 applies, where G is the conductance reflected to the high voltage side by tap changing [6].

2.3 Relation of Voltage Stability to Rotor Angle Stability

Voltage stability and rotor angle (or synchronous) stability are more or less interlinked. Transient voltage stability is often interlinked with transient rotor angle stability, and slower forms of voltage stability are interlinked with small-disturbance rotor angle stability. Often, the mechanisms are difficult to separate.

There are many cases, however, where one form of instability predominates. An IEEE report [9] points out the extreme situations: (a) a remote synchronous generator connected by transmission lines to a large system (pure angle stability—the one machine to an infinite bus problem) and (b) a synchronous generator or large system connected by transmission lines to an asynchronous load (pure voltage stability). Figure 2-5 shows these extremes.

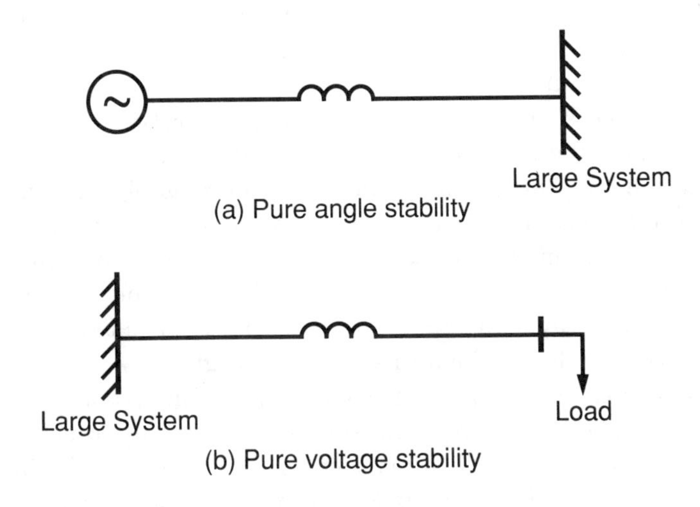

Fig. 2-5. Simple examples showing extreme situations.

Rotor angle stability, as well as voltage stability, is affected by reactive power control. In particular, small-disturbance ("steady-state") instability involving aperiodically increasing angles was a major problem before continuously-acting generator automatic voltage regulators became available. We can now see a connection between small-disturbance angle stability and longer-term voltage stability: generator current limiting (say by an overexcitation limiter) prevents normal automatic voltage regulation. Generator current limiting is very detrimental to both forms of stability.

Voltage stability is concerned with load areas and load characteristics. For rotor angle stability, we are often concerned with integrating remote power plants to a large system over long transmission lines. Voltage stability is basically *load stability*, and rotor angle stability is basically *generator stability*.

In a large interconnected system, voltage collapse of a load area is possible without loss of synchronism of any generators.

Transient voltage stability is usually closely associated with transient rotor angle stability. Longer-term voltage stability is less interlinked with rotor angle stability.

We can say that if voltage collapses at a point in a transmission system remote from loads, it's an angle instability problem. If voltage collapses in a load area, it's probably mainly a voltage instability problem.

2.4 Voltage Instability in Mature Power Systems

Voltage problems are expected in developing power systems. Likewise, voltage problems are expected following major system breakups. But why the recent concern in mature power systems?

One reason is intensive use of existing generation and transmission. This is because of difficulties in building new generation in load areas, and difficulties in building transmission lines from remotely-sited generation.

A second reason is increased use of shunt capacitor banks for reactive power compensation. Excessive use of shunt capacitor banks, while extending transfer limits, results in a voltage collapse-prone (brittle or fragile) network. Shunt capacitor bank reactive power output decreases by the square of voltage, hence the terms brittle or fragile.

Fast fault clearing, high performance excitation systems, power system stabilizers, and other controls are effective in removing transient stability-imposed transfer limits. With transient stability-imposed limits removed, either thermal capacity or voltage stability may dictate the transfer limits. An example provides insight into how voltage instability can become a problem in mature systems.

Example 2-1. Figure 2-6 shows a five-line 500-kV transmission network. Transient stability is not a problem. What about thermal limits? Outage of one line requires the remaining lines to pick up only 25% of the power of the out-of-service line. Long EHV lines are typically loaded below one and one-half times surge impedance loading. Thermal limits are typically about three times surge impedance loading. Unlike developing two- or three-line transmission networks, thermal limits will seldom be limiting and transmission can be highly utilized.

Now consider the effect of the five-line transmission system on voltage stability. We are fighting a nonlinear current-squared relation (I^2X series reactive power loss). Near surge impedance loading, the current in each line is 1000 amperes; a line outage will increase series reactive losses from 1200 MVAr (5 lines $\times$ 3 phases $\times$ 1000^2 $\times$ 80 ohms) to 1500 MVAr—an increase of 300 MVAr. Now consider several years load growth resulting in

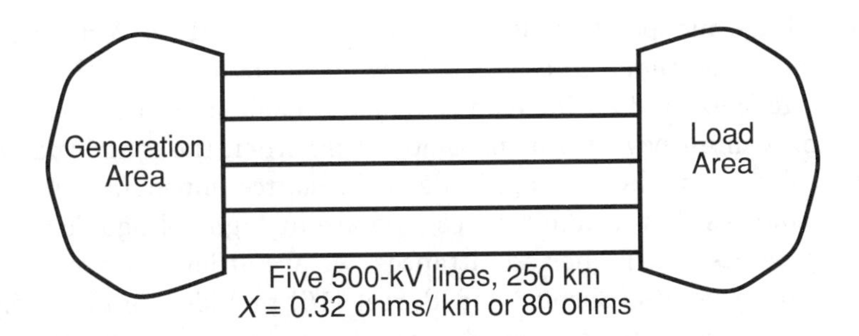

Fig. 2-6. Five-line transmission system representing a mature power system.

high utilization. With 1500 ampere loading, an outage increases series reactive losses from 2700 MVAr to 3375 MVAr—an increase of 675 MVAr or 225% over the surge impedance loading case. You can make similar calculations for two line outages. The effects of voltage drops, which increase series reactive losses (Equation 1.11) and reduce reactive power generation from transmission line capacitance, are not included in these calculations; these effects make the situation even worse. Because of these nonlinear effects, voltage stability problems may develop over a period of only a few years.

2.5 Introduction to Voltage Stability Analysis: *P–V* Curves

The slower forms of voltage instability are often analyzed as steady-state problems; power flow simulation is the primary study method. "Snapshots" in time following an outage or during load buildup are simulated. Besides these post-disturbance power flows, two other power flow based methods are widely used: *P–V* curves and *V–Q* curves. These two methods determine steady-state loadability limits which are related to voltage stability. Conventional power flow programs can be used for approximate analysis.

P–V curves are useful for conceptual analysis of voltage stability and for study of radial systems. The method is also used for large meshed networks where P is the total load in an area and V is the voltage at a critical or representative bus. P can also be the power transfer across a transmission interface or interconnection. Voltage at several busses can be plotted. A disadvantage is that the power flow simulation will diverge near the nose or maximum power point on the curve. Another disadvantage is that generation must be realistically rescheduled as the area load is increased.

For conceptual analysis, *P–V* curves are convenient when load characteristics as a function of voltage are analyzed. For example a resistive load can be plotted with $P_{\text{load}} = V^2/R$. The opposite extreme of a constant power (voltage independent) load* is even simpler—it's a vertical line on the *P–V* curve. Section 2.7 describes these ideas further.

First, let's expand on impedance loads. A basic network theorem tells us that maximum power transmission occurs when the magnitude of the load impedance equals the magnitude of the source impedance. For higher load impedances (lower admittances), we are at high voltage, low current operating points. For higher admittances, we are at low voltage, high current operating points. Barbier and Barret [16] provide the mathematical relations. For the simplest case of a resistance load and a reactance network, Figure 2-7 shows the relations of voltage, current, and power. As stated, maximum power occurs when the source and load impedance magnitudes are equal. We call the voltage at maximum power the critical voltage.

Example 2-2. For the simple thévenin system of Figure 2-7, find an expression for $P = f(V)$. For unity power factor load, determine the maximum power and the voltage at maximum power (critical voltage). Normalize the variables based on the short circuit power, E^2/X, with:

$$p = \frac{PX}{E^2},\ q = \frac{QX}{E^2},\ v = \frac{V}{E}$$

Solution: The relations from Chapter 1 are rewritten in normalized form.

$$P = \frac{EV}{X}\sin\delta,\ p = v\sin\delta$$

$$Q = \frac{EV\cos\delta}{X} - \frac{V^2}{X},\ q = v\cos\delta - v^2$$

Using the trigonometric identity $v^2\sin^2\delta + v^2\cos^2\delta = v^2$:

$$p = \sqrt{v^2 - v^2\cos^2\delta},\ \text{or}$$

$$p = \sqrt{v^2 - (q + v^2)^2}$$

At unity power factor, $p = \sqrt{v^2 - v^4}$; taking the derivative and setting it equal to zero, we get the critical voltage and maximum power.

$$\frac{dp}{dv} = \frac{1}{2}(v^2 - v^4)^{-1/2}(2v - 4v^3) = 0,\ 2v^2 = 1$$

*A constant power static load is non-physical and should be used with caution [15].

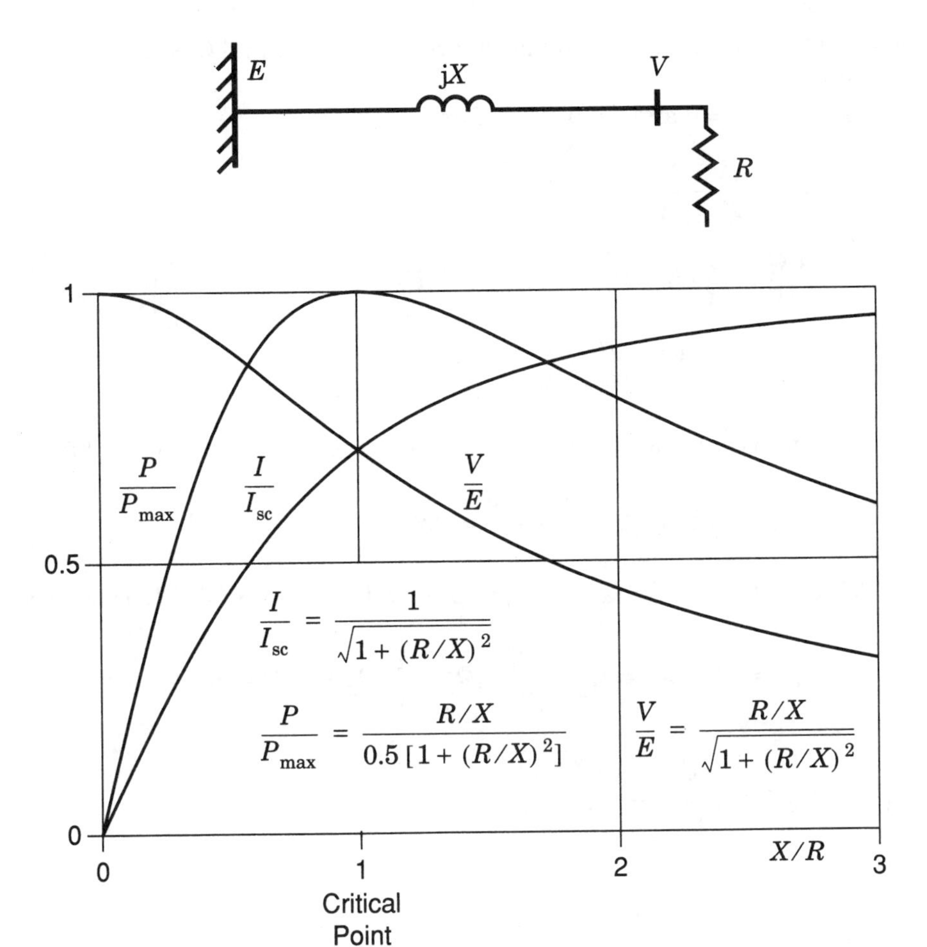

Fig. 2-7. Relations of voltage, current, and power for resistance load and reactance network. $I_{sc} = E/X$ is short circuit current. P_{max} and V_{crit} are calculated in Example 2-2.

$$v_{crit} = 1/\sqrt{2} = 0.707 \text{ and } p_{max} = \sqrt{1/2 - 1/4} = 0.5$$

Also, at maximum power, $\delta = \sin^{-1}(p_{max}/v_{crit}) = 45°$.

For the case of resistive load, we can verify that maximum power occurs when the load resistance R equals the source reactance X:

$$P_{max} = \frac{p_{max}E^2}{X} = \frac{0.5\,(\sqrt{2}V_{crit})^2}{X} = \frac{V_{crit}^2}{X} = \frac{V^2}{R}$$

Example 2-3. Repeat the previous problem with a purely reactive load. Calculate the "Voltage Collapse Proximity Indicator," ($VCPI = dQ_g/dQ$)

where Q_g is generated or sending end reactive power and Q is load reactive power [17].

Solution: $P = 0$ and $\delta = 0$. Therefore:

$$Q = \frac{EV}{X} - \frac{V^2}{X}$$

$$\frac{dQ}{dV} = \frac{1}{X}(E - 2V) = 0, \; V_{crit} = E/2$$

$$Q_{max} = \frac{2V^2}{X} - \frac{V^2}{X} = \frac{V^2}{X} = \frac{V^2}{X_{load}}$$

We again confirmed the maximum power theorem. In normalized form:

$$v_{crit} = 0.5, \; q_{max} = \frac{Q_{max}X}{E^2} = 0.25.$$

We calculate the voltage collapse proximity indicator (*VCPI*) as follows:

$$Q_g = Q + XI^2 = Q + \frac{XQ_g^2}{E^2}, \; Q_g^2 - \frac{E^2}{X}Q_g + \frac{E^2}{X}Q = 0$$

$$Q_g = \frac{1}{2}\left[\frac{E^2}{X} \pm E\sqrt{\frac{E^2}{X^2} - \frac{4Q}{X}}\right]$$

$$\frac{dQ_g}{dQ} = \frac{1}{\sqrt{1 - \frac{4XQ}{E^2}}} = \frac{1}{\sqrt{1 - \frac{Q}{Q_{max}}}}$$

The voltage, V, goes from E at no load to $E/2$ at maximum load (Q_{max}). What about the *VCPI*? It goes from unity at no load to infinity at maximum load. Near maximum load, extremely large amounts of reactive power are required at the sending end to support an incremental increase in load. The *VCPI* is thus a very sensitive indicator of impending voltage collapse. The related quantities, reactive reserve activation and reactive losses, are also sensitive indicators.

For the elementary model, Figure 2-8 shows the family of normalized *P–V* curves for different power factors. At more leading power factors the maximum power is higher (leading power factor is obtained by shunt compensation). The critical voltage is also higher, which is a very important aspect of voltage stability.

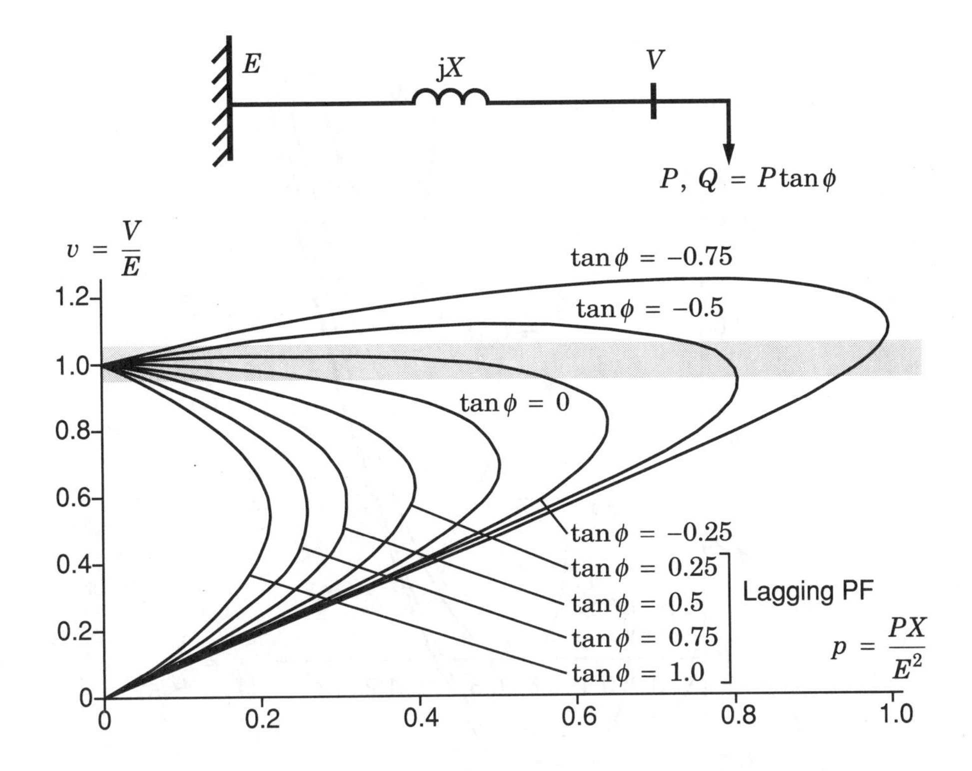

Fig. 2-8. Normalized P–V curves for fixed (infinite) source and reactance network [1]. Corresponding power factors for tan ϕ = 1.0, 0.75, 0.5, 0.25, and 0 are 0.707, 0.8, 0.894, 0.97, and 1.0.

2.6 Introduction to Voltage Stability Analysis: *V–Q* Curves

First, we can map the normalized p–v curves shown on Figure 2-8 onto v–q curves. For constant values of p, we note the q and v values (two pairs for each power factor), and then replot. Figure 2-9 shows the result. Again, the critical voltage is very high for high loadings (v is above 1 pu for $p = 1$ pu). The right side represents normal conditions where applying a capacitor bank raises voltage. The steep-sloped linear portions of the right side of the curves are equivalent to Figure 1-8 (rotate Figure 1-8 clockwise 90°).

For large systems, the curves are obtained by a series of power flow simulations. *V–Q* curves plot voltage at a test or critical bus versus reactive power on the same bus. A fictitious synchronous condenser is represented at the test bus. In computer program parlance, the bus is converted to a "PV bus" without reactive power limits. Power flow is simulated for a series

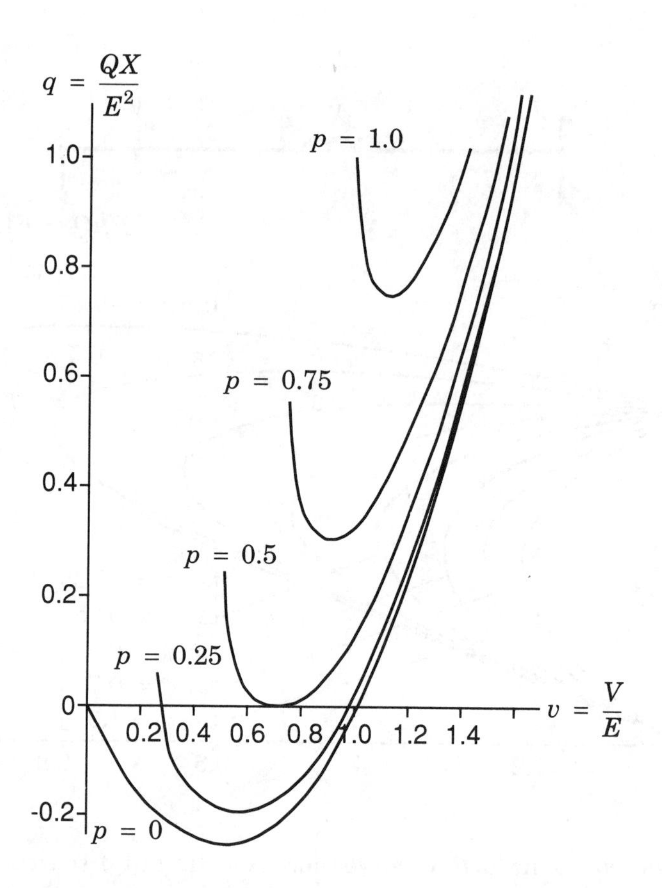

Fig. 2-9. Normalized v–q curves for fixed (infinite) source and reactance network [1]. Loads are constant power.

of synchronous condenser voltage schedules, and the condenser reactive output is plotted versus scheduled voltage. Voltage is the independent variable and is the abscissa variable. Capacitive reactive power is plotted in the positive vertical direction. Without application of shunt reactive compensation at the test bus, the operating point is at the zero reactive point—corresponding to removal of the fictitious synchronous condenser.

(These curves are often termed *Q–V* rather than a *V–Q* curves, but the *V–Q* terminology stresses that voltage rather than reactive power load is the independent variable. *Q–V* curves are produced by scheduling reactive load rather than voltage.)

V–Q curves have several advantages:

- Voltage security is closely related to reactive power, and a *V–Q* curve gives reactive power margin at the test bus. The reactive power margin is the MVAr distance from the operating point to either the bottom of the curve, or to a point where the voltage squared characteristic of an applied capacitor is tangent to the *V–Q* curve (Figure 2-10). The test bus could be representative of all busses in a "voltage control area" (an area where voltage magnitude changes are coherent).

- *V–Q* curves can be computed at points along a *P–V* curve to test system robustness.
- Characteristics of test bus shunt reactive compensation (capacitor, SVC, or synchronous condenser) can be plotted directly on the *V–Q* curve. The operating point is the intersection of the *V–Q* system characteristic and the reactive compensation characteristic (Figure 2-10b). This is useful since reactive compensation is often a solution to voltage stability problems.
- The slope of the *V–Q* curve indicates the stiffness of the test bus (the ΔV for a ΔQ).
- For more insight, the reactive power of generators can be plotted on the same graph. When nearby generators reach their VAr limits, the slope of the *V–Q* curve becomes less steep and the bottom of the curve is approached.

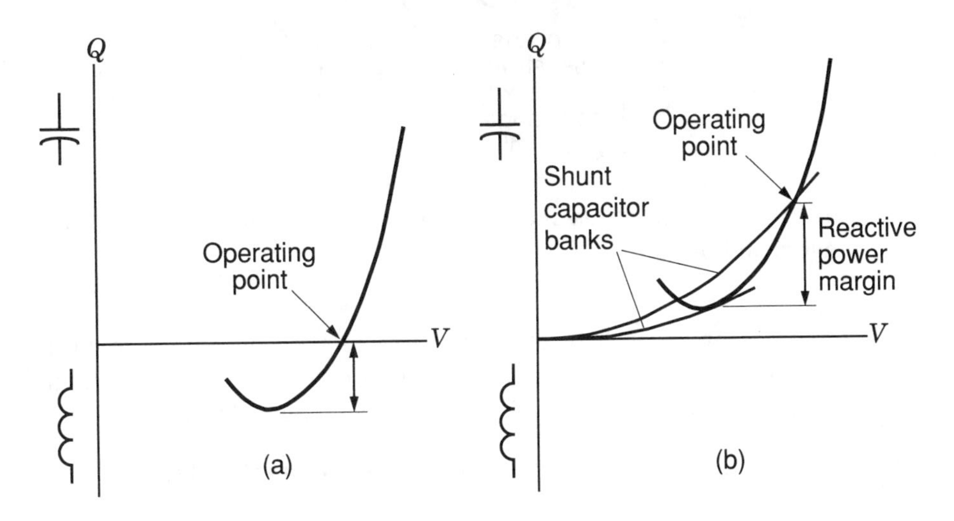

Fig. 2-10. Reactive power margins.

From a computation viewpoint, the artificial PV bus minimizes power flow divergence problems. Solutions can be obtained on the left side of the curve—divergence only occurs when voltages at busses away from the PV bus are dragged down. Generation rescheduling needs are minimal since the only changes in real power are caused by changes in losses. Starting values from the previous solution at a slightly different scheduled voltage are used so that each power flow solution is fast. The process can be automated so that the entire curve is computed at one time.

The effect of voltage sensitive loads, or of tap changing reaching limits, can be shown on *V-Q* curves. *V-Q* curves with voltage sensitive loads (i.e., prior to tap changing) will have much greater reactive power margins and much lower critical voltages. When tap changers hit limits, the curves tend to flatten out rather than turn up on the left side. These ideas are sketched on Figure 2-11.

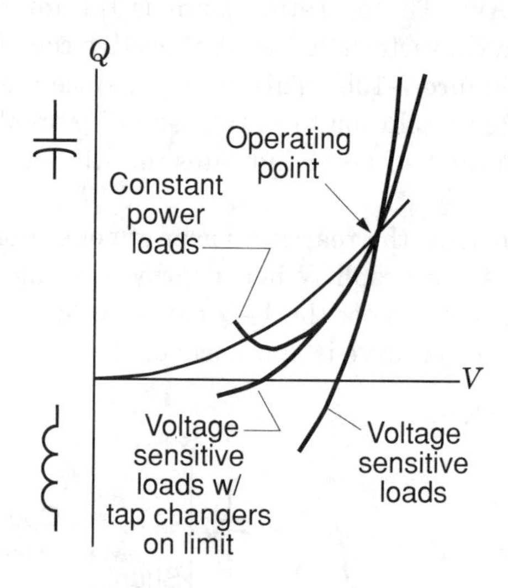

Fig. 2-11. *V–Q* curve sketches showing effect of voltage sensitive loads and tap changers on limit.

V–Q curves are presently the workhorse method of voltage stability analysis at many utilities. Since the method artificially stresses a single bus, conclusions should be confirmed by more realistic methods.

2.7 Graphical Explanation of Longer-Term Voltage Stability

Using *P-V* curves, we examine voltage stability as related to load characteristics. We include the effects of tap changing, constant energy loads, and generator current limiting.

Figure 2-12 shows a conceptual system model. Over a voltage range of about 0.9–1.1 per unit, we approximate induction motor load as constant power static load and assume motor reactive power demand is matched by shunt compensation. Different proportions of constant and resistive load are assumed. (Over a small voltage range, other load components such as

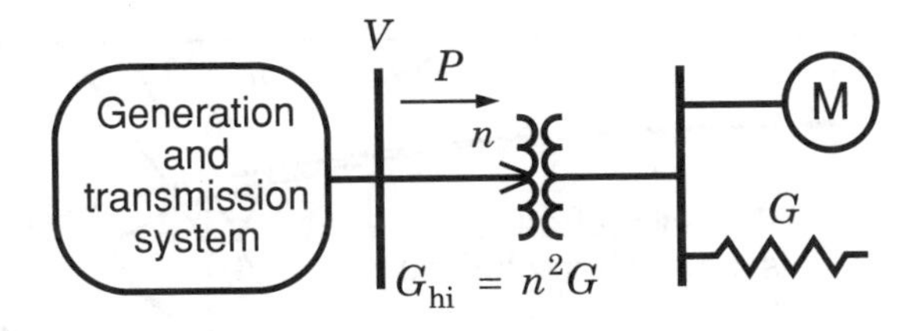

Fig. 2-12. Conceptual model.

lighting can be modeled as combinations of constant and resistive load.) The resistance (or conductance, G) is reflected to the transmission side of the LTC transformer by the square of the turns ratio, n. Typical tap changer range is ±10%.

We examine two extremes of load composition. First, we assume the load is 75% motor and 25% resistive—this approximates a load area that is primarily industrial, or a load area that has large amount of air conditioning. Then we assume the load is 25% motor and 75% resistive—this approximates a load area with a large amount of electric space heating.

High proportion of motor load. Figure 2-13 shows three system characteristics, plus load characteristics for the 75% motor load case. The operating point is an intersection of the system and load characteristics.

The three system characteristics are: (1) pre-disturbance, (2) post-disturbance before generator current limiting, and (3) post-disturbance with current limiting on some generators [16,18]. Generator field current limiting in response to overload is by overexcitation limiters or by operator intervention. Each system characteristic has a maximum power point representing the steady-state power transfer capability of the generation and transmission system. The system characteristics are purely conceptual.

Also shown on Figure 2-13 are load characteristics for the 75% motor load case. Two load characteristics are shown: the initial load characteristic and the characteristic with +10% tap changing. The load equation is $P = 0.75 + 0.25n^2GV^2$. Following the disturbance, the voltage will drop and the load characteristic will intersect one of the post-disturbance system characteristics. The drop in voltage will reduce the resistive load until tap changer regulation of the low side voltage. If there is no tap changing, or if tap changers are at limits, load conductance will increase for thermostatically-controlled loads or for other constant energy diversified loads.

Figure 2-13 shows that the load characteristics are quite unfavorable. With tap changing and generator current limiting, the operating point is

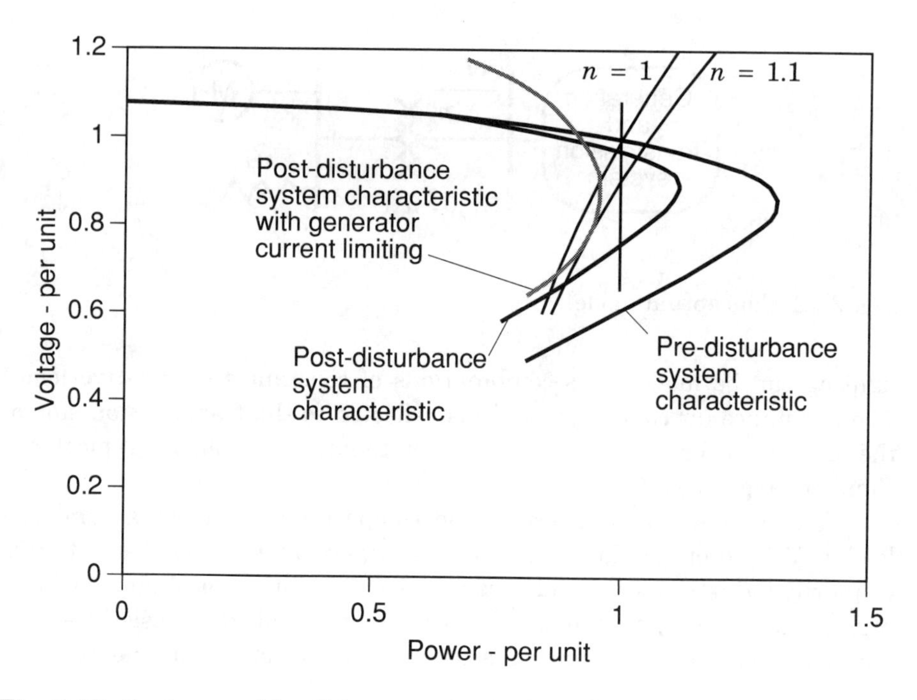

Fig. 2-13. System and load characteristics for 75% motor load case. Load characteristic: $P = 0.75 + 0.25n^2GV^2$ (G = 1 pu.) Initial conditions are one per unit voltage and power.

lost (ill-defined or no intersection of system and load characteristics). Loss of an operating point starts voltage collapse. Dynamic simulation with accurate models (including possible ac contactor dropout) is required to determine performance. In fact, motor instability will occur before instability predicted by steady state analysis [6,19].* If the disturbance involves a severe short circuit, motors may not reaccelerate. Nevertheless, Figure 2-13 has conceptual value.

High proportion of resistive load. Figure 2-14 shows corresponding curves for the 75% resistive load case (the system characteristics are unchanged). The load equation is $P = 0.25 + 0.75n^2GV^2$, where conductance, G, is one per unit for curves a and b, and 1.2 per unit for curve c. The additional conductance for curve c could be from thermostatic regulation of part of the resistive load and could represent all heaters on. For LTC and

*Chapter 6 describes the effects. The point of constant power is actually internal to the motor and the motor impedance should be considered.

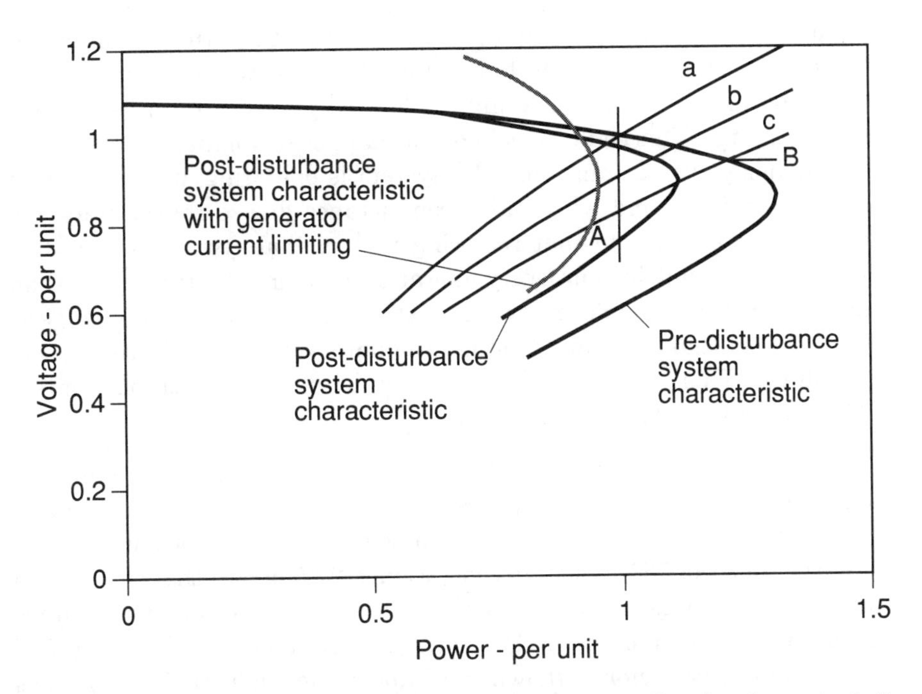

Fig. 2-14. System and load characteristics. Load characteristic: $P = 0.25 + 0.75n^2GV^2$. Load curve a is for $n = 1$; Curve b is for $n = 1.1$. Curve c is for 20% more conductance ($G = 1.2$) and $n = 1.1$.

thermostatic regulation, the load will not increase above the initial one per unit value (vertical line).

Without generator current limiting, the post-disturbance operating point is at the intersection of the vertical line and the post-disturbance system characteristic. The load is fully restored and the tap ratio, n, is approximately 1.03. With current limiting but without added conductance, the operating point is at the intersection of curve b and the gray curve. With the added conductance, the operating point is Point A.

The load characteristics are more favorable and loss of operating point (loss of intersection) causing voltage collapse is unlikely. Stabilization at abnormally low voltage on the underside of the *P–V* curve is possible. In a realistic system, the outcome would depend on motor performance (including stalling or disconnection), protective relaying operations, and the amount of automatic or manually regulated constant energy load. If the operating point is on the underside side of a *P–V* curve, an increase in load conductance results in a reduction of load power. As described in Appendix F, power systems have actually stably operated at abnormally low voltage during a partial voltage collapse (Figure F-1).

Load curve b intersects the system characteristic with generator current limiting at its nose. Additional tap changing would decrease load power. This means the tap changing is lowering, rather than increasing, load-side voltage. This could be called tap changer instability.

A further point: Assume that the system is operating at Point A with taps at their limits and with added conductance due to thermostats. On Figure 2-14, we can note that restoring the original system (say by line reclosing) will result in temporary operation at Point B with higher than initial power and possible overvoltage at the load side of tap changers. (This assumes that generator voltage regulators immediately respond to high voltage. This is true for continuously acting overexcitation limiters that do not trip regulators to manual.)

Effect of shunt capacitor bank switching. Referring to the family of *P–V* curves for different power factors shown on Figure 2-8, we can analyze the effect of switching on a capacitor bank while on the bottom side of a *P–V* curve. For the 75% resistive case, Figure 2-15 shows that inserting a capacitor bank moves the operating point from Point A to Point B. Voltage and power are both increased. The voltage/reactive power performance is normal. Similar behavior is shown in Chapter 4 for induction motor loads (Example 4-4).

Both the stable operation on the bottom side of a *P–V* curve and the capacitor switching effects have been verified by time domain simulation [20]. Pal [6] and Van Cutsem [18] describe some of these effects with mathematical rigor.

2.8 Summary

We have described two types of voltage stability: transient voltage stability and longer-term voltage stability. Longer-term voltage stability involving loads that are inherently voltage sensitive has been of greatest interest in recent years.

Voltage stability involves the load, transmission, and generation subsystems of large power systems. Three key aspects of voltage stability are:

1. the load characteristics as seen from the bulk power network;
2. the available means for voltage control at generators and in the network; and
3. the ability of the network to transfer power, particularly reactive power, from the point of production to the point of consumption.

Voltage instability and collapse is a dynamic phenomena and normally a large disturbance phenomena. The network steady-state loadability limit

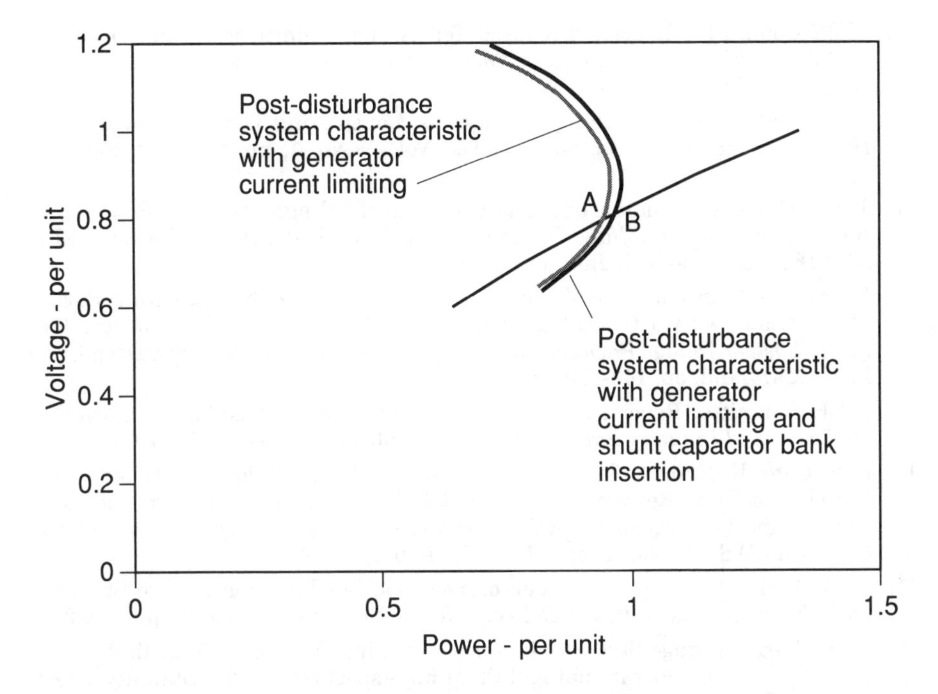

Fig. 2-15. Effect of capacitor bank insertion on bottom side of *P–V* curve with highly voltage sensitive load. Load characteristic same as curve c of Figure 2-14.

(maximum power point on a *P–V* curve) is not necessarily the voltage instability limit. Nevertheless, static, power flow based analysis of the post-disturbance steady-state is often a useful method of analysis for longer-term voltage stability.

In later chapters we analyze voltage stability in more depth. First, however, we examine the characteristics of the various equipment influencing voltage stability.

References

1. CIGRÉ Working Group 38.01, "Planning Against Voltage Collapse," *Electra*, pp. 55–75, March 1987.
2. B. M. Weedy, *Electric Power Systems*, Third Edition Revised, John Wiley & Sons, 1987 (earlier editions 1967, 1972, and 1979).
3. CIGRÉ Task Force 38-02-10, *Modelling of Voltage Collapse Including Dynamic Phenomena*, 1993.
4. D. J. Hill, Per-Anders Löf, and G. Andersson, "Analysis of Long-Term Voltage Stability," *Tenth Power System Computing Conference*, pp. 1252–1259, Graz Austria, 1990.

5. IEEE Committee Report, "Proposed Terms and Definitions for Power System Stability," *IEEE Transactions on Power Apparatus and Systems*, Vol. PAS-101, No. 7, pp. 1894–1898, July 1982.
6. M. K. Pal, "Voltage Stability Conditions Considering Load Characteristics," *IEEE Transactions on Power Systems*, Vol. 7, No. 1, pp. 243–249, February 1992.
7. T. Van Cutsem, "Dynamic and Static Aspects of Voltage Collapse," *Proceedings: Bulk Power System Voltage Phenomena—Voltage Stability and Security*, EPRI EL-6183, pp. 6-55–6-79, January 1989.
8. Y. Tamura, "A Scenario of Voltage Collapse in a Power System with Induction Motor Loads with a Cascaded Transition of Bifurcations," *Proceedings: Bulk Power System Voltage Phenomena II: Voltage Stability and Security*, Deep Creek Lake, Maryland, pp. 143–146, 4–7 August 1991.
9. IEEE Committee Report, *Voltage Stability of Power Systems: Concepts, Analytical Tools, and Industry Experience*, IEEE publication 90TH0358-2-PWR.
10. E. S. Cate, K. Hemmaplardh, J. W. Manke, and D. P. Gelopulos, "Time Frame Notion and Time Response of the Models in Transient, Mid-Term and Long-Term Stability Programs," *IEEE Transactions on Power Apparatus and Systems*, Vol. PAS-103, No. 1, pp. 143–151, January 1984.
11. C. W. Taylor, "Concepts of Undervoltage Load Shedding for Voltage Stability," *IEEE Transactions on Power Delivery*, Vol. 7, No. 2, pp. 480–488, April 1992.
12. H. K. Clark, "Voltage Stability: Criteria, Planning Tools, Load Modeling," EPRI/NERC Forum on Operational and Planning Aspects of Voltage Stability, Breckenridge, Colorado, 14–15 September 1992.
13. D. McInnis, "South Florida Blackout," unpublished Florida Power & Light report.
14. CIGRÉ Working Group14-07 and IEEE Working Group15.05.05, *Guide for Planning DC Links Terminating at AC Systems Locations Having Low Short-Circuit Capacities, Part I: AC/DC Interaction Phenomena*, CIGRÉ, June 1992.
15. M. K. Pal, discussion of "An Investigation of Voltage Instability Problems," by N. Yorino et al., *IEEE Transactions on Power Systems*, Vol. 7, No. 2, pp. 600–611, May 1992.
16. C. Barbier and J.-P. Barret, "Analysis of Phenomena of Voltage Collapse on a Transmission System," *Revue Generale de l'electricite*, Vol. 89, October 1980, pp. 672-690.
17. J. Carpentier, R. Girard, and E. Scano, "Voltage Collapse Proximity Indicators Computed from an Optimal Power Flow," *Proceedings of the 8th Power System Computing Conference*, pp. 671–678, Helsinki, 1984.
18. T. Van Custsem, "Voltage Collapse Mechanisms: A Case Study," *Proceedings: Bulk Power System Voltage Phenomena II: Voltage Stability and Security*, Deep Creek Lake, Maryland, pp. 85–101, 4–7 August 1991.
19. R. J. Thomas and A. Tiranuchit, "Dynamic Voltage Instability," *Proceedings of the 26th Conference on Decision and Control*, Los Angeles, pp. 53–58, December 1987.
20. C. W. Taylor, "A Conceptual Analysis of Voltage Stability as Related to Load Characteristics," *Survey of Voltage Collapse Phenomena*, North American Electric Reliability Council, 1991.

Transmission System Reactive Power Compensation and Control

It's what you learn after you know it all that counts.
John Wooden*

Reactive power compensation is often the most effective way to improve both power transfer capability and voltage stability. In this chapter we describe transmission system compensation; in the next chapter we describe distribution system compensation.

Reactive power compensation can be divided into series and shunt compensation. It can also be divided into active and passive compensation; active compensation means a feedback control system regulates voltage or other variables. Common forms of reactive compensation are series capacitor banks, shunt reactors and capacitor banks, and static var compensators. Under-load transformer tap changing also provides voltage/reactive power control.

First we review transmission line characteristics.

3.1 Transmission System Characteristics

Chapter 1 introduced the characteristics of reactive power transmission. When a long line is heavily loaded, reactive power cannot transmitted even with large voltage gradients. Now we probe deeper and answer some questions. Why and when are transmission lines compensated? What parameters need compensation? How do transmission line parameters vary with voltage class and line design? When should compensation be controlled?

Bulk power transmission is generally at 230-kV and higher voltages. Voltages above 230-kV are termed extra high voltage (EHV). Bulk power transmission lines are often quite long and are heavily loaded during peak

---

*John Wooden was a very successful basketball coach at the University of California at Los Angeles.

loads. They may be lightly loaded off peak. Two of the characteristics of EHV overhead transmission lines are low losses and multiple subconductors per phase (bundled conductors).

A fundamental aspect of reactive power compensation and control is reactive power balance. Transmission lines both produce and consume reactive power, and the net values must be either absorbed or generated by the system at each line terminal. Transmission line shunt capacitance (or "charging") produces reactive power proportional to the square of the voltage. Since the voltage must be kept within about ±5% of nominal voltage, the reactive power production is relatively constant. Transmission line series inductance consumes reactive power proportional to the square of the current. Since the current varies from heavy load periods to light load periods, the transmission line reactive consumption varies. Therefore the net transmission line reactive power varies over the load cycle. We can state:

transmission line production = V^2B (relatively constant),
transmission line consumption = I^2X (variable).

$B = \omega C$ is the line shunt susceptance and $X = \omega L$ is the line series reactance.

Surge impedance loading. We are often interested in the loading where production equals consumption. This is called the natural or surge impedance loading. For an incremental length of line of reactance x and susceptance b, we set $V^2b = I^2x$, and solve for the surge or characteristic impedance:

$$Z_0 = \frac{V}{I} = \sqrt{\frac{x}{b}} = \sqrt{\frac{l}{c}} \qquad (3.1)$$

The natural or surge impedance loading is then:

$$P_0 = \frac{V^2}{Z_0} \qquad (3.2)$$

Surge impedance loading is, in many respects, the ideal loading. Not only is reactive power production equal to reactive power consumption, but the voltage and current profiles are uniform along the line. The uniform (flat) voltage profile is especially desirable since voltage can be held near the maximum value. The voltages and currents are also in phase at every point along the line.

Long lines cannot be loaded much above the uncompensated surge impedance loading. References 1 and 2 develop a "loadability curve" shown on Figure 3-1. To insure small disturbance angle stability and for other reasons, lines longer than about 450 km need to be compensated.

We will discuss several other reasons for compensation. But first we will review transmission line parameters.

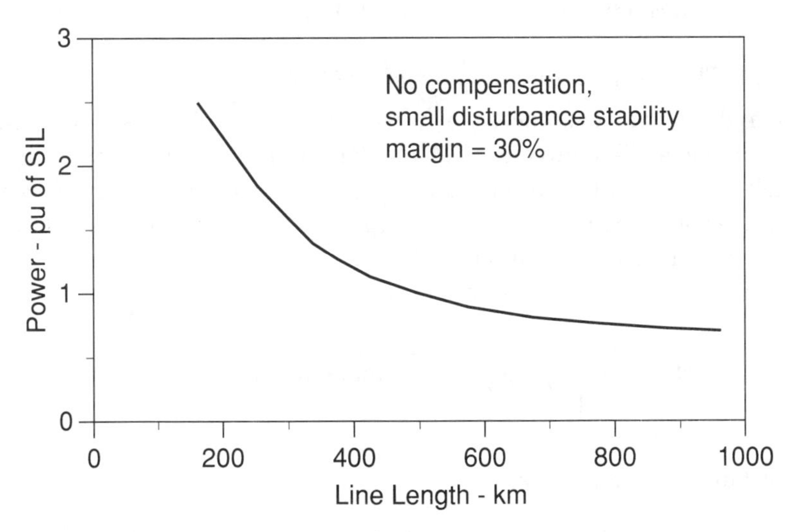

Fig. 3-1. Loadability curve [2]. Voltage drop limits loadability for lines shorter than 320 km (200 miles). For longer lines, small disturbance stability margins limit loadability.

Transmission line parameters. Let's review some basics. The most important parameters are series resistance and reactance, and shunt susceptance. Series resistance affects losses and loadability (thermal and sag limits). Resistance of high voltage and EHV lines is small, however, and can often be ignored. For EHV lines, *positive sequence X/R* ratios are 10–20.

An equation for inductive reactance is [3]:

$$x = \omega l = 2\omega 10^{-4} \ln \frac{GMD}{GMR} \quad \Omega/\text{km} \tag{3.3}$$

where ω is the power system radian frequency. GMD is the geometric mean distance between phases: $GMD = (d_{ab} + d_{ac} + d_{bc})^{1/3}$. GMR is the geometric mean radius and can be obtained from conductor tables; $GMR \cong 0.8r$ where r is the conductor radius.

For bundled conductors (several subconductors per phase) with spacing s between adjacent subconductors, the equivalent GMR is [3]:

$$GMR_{\text{equiv}} = \left[n \times GMR \left[\frac{s}{2\sin\pi/n} \right]^{n-1} \right]^{1/n} \tag{3.4}$$

For two and three conductor bundles, the equivalent *GMR*s are:

Two conductors $\sqrt{s \times GMR}$

Three conductors $\sqrt[3]{s^2 \times GMR}$.

To reduce the reactance, we must reduce the phase spacing (*GMD*) and/or increase the equivalent *GMR*. *GMD* is reduced by compact design and delta phase configuration. GMR_{equiv} is reduced mainly by increasing the number of subconductors; for example, 500-kV lines are built with two, three, and four subconductors per phase.

A corresponding equation for shunt susceptance is [3]:

$$b = \omega c = \frac{\omega 10^{-6}}{18 \ln \dfrac{GMD}{r}} \text{ S/km (siemens/km)} \tag{3.5}$$

For bundled conductors:

$$r_{\text{equiv}} = \left[n \times r \left[\frac{s}{2\sin\pi/n} \right]^{n-1} \right]^{1/n} \tag{3.6}$$

The charging reactive power is:

$$Q_{\text{chg}} = V^2 b \tag{3.7}$$

Reduced phase spacing and bundled conductors reduce line inductance and reactance, and increases line capacitance and susceptance. This increases the surge impedance loading (Equations 3.1 and 3.2) and effective transmission capability. The increased reactive power generation due to higher capacitance causes problems at light loads, and long EHV lines usually need shunt reactor compensation. The reactors are sometimes switched off during heavy load.

Although bundled subconductors substantially increases the equivalent radius of phase conductors, the decrease in reactance and the increase in susceptance are modest because of the logarithmic functions in Equations 3.3 and 3.5.

Figure 3-2 shows a high capacity, double-circuit 500-kV transmission line using three subconductors per phase.

Example 3-1. A 500-kV line has phase conductors in a delta configuration with 9.4 meter spacing. The conductor radius is 2.54 cm (1 inch), and the

Fig. 3-2. 500-kV double-circuit transmission line. *Bonneville Power Administration.*

GMR is 2.03 cm. Spacing between subconductors is 45.72 cm (18 inches). Frequency is 60 Hz. Compare the inductive reactance, charging reactive power, surge impedance, and surge impedance loading for one, two, three, and four conductors per phase. For the charging power and the surge impedance loading calculation, use 525 kV average operating voltage.

Solution: Using Equations 3.1 through 3.4 the calculations are straight-forward. Table 3-1 summarizes the results. Compared with two subconductors, four subconductors increases surge impedance loading by 23%. This is equivalent to 33% uniformly distributed series compensation (Equation 3.8).

Transmission line theory. Lines longer than about 400 km require special analysis. Starting with a distributed parameter model, the "wave equation" is solved for voltages and currents along a transmission line. For sinusoidal steady-state analysis, we eliminate the time parameter and are left with equations as a function of distance. For lossless lines, Miller [4] covers the subject very well. Figure 3-3 shows a transmission line model for a lossless line.

Table 3-1

#Subcond.	x ohms/km	Q_{chg} MVAr/km	Z_0 ohms	P_0 MW
1	0.46	0.98	361	762
2	0.35	1.29	271	1015
3	0.31	1.45	241	1142
4	0.28	1.58	221	1246

For lines shorter than about 200 km, small angle approximations can be made for the trigonometric corrections terms. The correction terms are then unity and the well-known lumped parameter pi model results. For power flow and stability computer programs, longer lines can be broken into lumped parameter sections of about 150–200 km. This is an alternative to use of the correction terms. For lossy lines the correction terms involve hyperbolic functions.

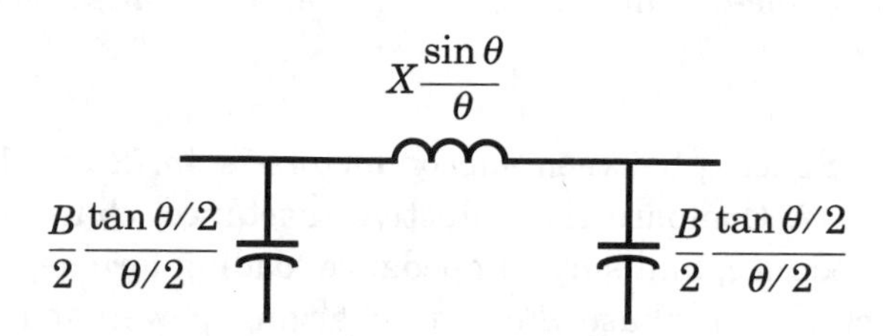

Fig. 3-3. Pi transmission line model with long-line correction terms. $X = ax$ and $B = ab$, where a is line length, and x and b are reactance and susceptance per unit length. θ is line length and equals $a\sqrt{xb}$ radians.

Example 3-2. A 200 km Bonneville Power Administration 500-kV line (3-1192 MCM Bunting conductors) has parameters $x = 0.32$ Ω/km and $b = 5.02$ μS/km. Calculate the correction terms for the equivalent circuit of Figure 3-1.

Solution: $\theta = a\sqrt{xb} = 0.253$ radians $= 14.5°$, then

$$\frac{\sin\theta}{\theta} = 0.991 \text{ and } \frac{\tan\theta/2}{\theta/2} = 1.0057$$

Both correction terms are close to unity. As an exercise, repeat the calculations for a 400 km line. Compare the resulting models with the series combination of two 200 km lines (use ABCD generalized circuit constants to calculate the series equivalent).

Example 3-3. Very long lines require distributed parameter analysis. For the line parameters of Example 3-3, calculate the voltage at the open receiving end of a quarter-wavelength line.

Solution: The wavelength for 60 Hz is: $\theta = a\sqrt{xb} = 2\pi$ or $a = 2\pi/\sqrt{xb} = 4957$ km. A quarter-wavelength is 1239 km or 770 miles. The following equation for voltage can be derived [3]: $V_s = V_r \cos\theta + jZ_0 I_r \sin\theta$. Since the receiving end is open, $I_r = 0$ and $V_r = V_s/\cos\theta$. V_r approaches infinity as θ approaches 90°! Obviously, shunt reactors are needed on very long lines.

Power system engineers are well-advised to memorize Equations 3.1, 3.2, and 3.7, plus typical parameters of several classes of transmission lines. Combinations of inductive reactance per unit length, surge impedance, surge impedance loading, and charging power should be immediately available for quick calculations. Table 3-2 suggests values to be memorized.

Susceptance and charging power can be calculated knowing inductive reactance and surge impedance. Using Table 3-2, we can calculate that b = 5.12 μS/km for a three subconductor line. At 500-kV, the surge impedance loading is $500^2/250$ or 1000 MW. The charging reactive power is about 1.3 MVAr/km, or about 2 MVAr/mile.

Table 3-2

# subconductors	Z_0 ohms	x ohms/km (ohms/mile)
one	400	0.5 (0.8)
two	300	0.037 (0.6)
three	250	0.32 (0.52)

Cables. Cable parameters are very different. Because of the close spacing, inductive reactance is lower and capacitance is higher. A 345-kV cable has inductive reactance of 0.09–0.16 Ω/km (0.15–0.26 Ω/mile) and charging reactive power of about 12 MVAr/km (20 MVAr/mile). Because of the high charging power, a key parameter of cables is the *critical length*, defined as the length at which the charging power equals the cable thermal capacity. For EHV cables, the critical length may be around 25 km.

3.2 Series Capacitors

Series capacitor compensation has traditionally been associated with long transmission lines and with improving transient stability. Nowadays, however, series capacitors are also applied on shorter lines to improve voltage stability.

Series compensation reduces net transmission line inductive reactance. The reactive generation (I^2X_c) compensates for the reactive consumption (I^2X) of the transmission line. Series capacitor reactive generation increases with the current squared, thus generating reactive power when most needed. This instantaneous, inherent self regulation is very valuable. Because of the self regulation, series compensation should be compared with active shunt compensation (static var compensators) rather than passive shunt capacitor banks.

At light load, series capacitors have little effect. Shunt reactors are needed for long lines.

Series capacitor equipment is mounted on a platform at line potential and includes the capacitors, spark gap protection, metal-oxide varistor (MOV), bypass switch, and control and protection. References 4–8 cover many aspects of series capacitor engineering. References 9–11 describe recent installations. References 12–14 cover the important subject of subsynchronous resonance. We concentrate on aspects closely related to voltage stability and recommend Miller [4] for discussions of reactive compensation related to very long lines and angle stability.

Figure 3-4 shows a single-line schematic of series compensation, and Figure 3-5 shows an overhead view of a series capacitor installation on a two-circuit 500-kV transmission system.

Series capacitors are almost always in transmission lines, rather than within a substation bus arrangement. They may be at line terminals or at intermediate points along the line. Except for very long lines, they are located at line terminals unless developed substations are available along the line. Compensated lines up to 75–80% are in operation or are planned; overall sending end to receiving end compensation is much less. For voltage profile or protective relaying reasons, compensation at one location must be limited. For 70% compensation, 35% compensation at each terminal is typical.

We are concerned with reactive power balance. Series compensation raises the effective surge impedance loading. (Shunt reactor compensation lowers the effective surge impedance loading.) If we approximate the lumped compensation with uniformly distributed compensation, we easily obtain the following formula for compensated surge impedance loading:

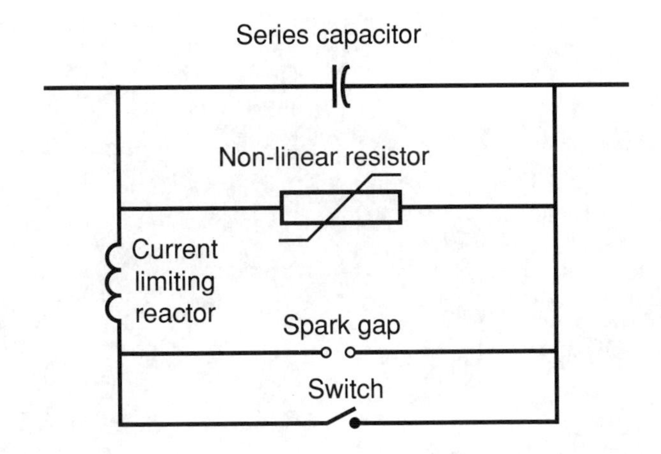

Fig. 3-4. Schematic of typical series capacitor compensation equipment.

$$P_0' = P_0\sqrt{\frac{1-k_{sh}}{1-k_{se}}} \tag{3.8}$$

where k_{se} is the degree of series compensation and k_{sh} is the degree of shunt compensation (positive for shunt reactors).

For example, for 50% series compensation and no shunt compensation, the surge impedance loading is increased by $\sqrt{2}$.

A difficulty with series compensation is overload for parallel line outages. For two parallel lines, and with an outage of one line, the current in the remaining line approximately doubles. The reactive power generated by series capacitors quadruples. Since the reactive power rating and the cost of series compensation is proportional to the current squared, advantage is taken of short-time overload capability. Standards allow overload current (capacitor overvoltage) of 135% for thirty minutes and 150% for five minutes. In voltage emergencies, the time-overload capability allows time for operators to reschedule generation, bring gas turbines on line, or shed load.

Example 3-4. For two series compensated parallel lines, calculate the MVAr ratings required for outage of one parallel line. Allow 150% voltage/-current overload for a short period.

Solution [5]: The capacitor bank must withstand twice normal voltage and current. Therefore (rated current) x (1.5 per unit) equals 2 per unit, and rated current equals 1.33 per unit of normal full load current. The bank MVAr rating must be increased by a factor of 1.33^2 or 1.78. For exam-

Fig. 3-5. Overhead view of 500-kV series capacitor platforms. *General Electric Company.*

ple, a capacitor bank with normal full load current of 1500 amperes would be rated 2000 amperes.

Another difficulty with series compensation is voltage profile during heavy load conditions accompanied by transmission outages. The voltage on one or the other side of the series capacitor will probably be too high. These very high current conditions are expected during voltage emergencies. Furthermore, transmission voltages should be kept as high as possible to minimize reactive losses (Equation 1.11). The phasor diagrams of Figure 3-6 show the problem; for the high current condition, we must somehow control the voltage-current power factor angles. Solutions involve smaller amounts of series compensation at a particular location and judicious use of shunt reactors.

Series capacitors can be normally bypassed but rapidly inserted following large disturbances. This practice has been used on the northern part of the Pacific AC Intertie for many years [8]. Voltage stability can be improved if, owing to the bypassed series capacitors, more shunt capacitors are applied, keeping generator power factor close to unity.

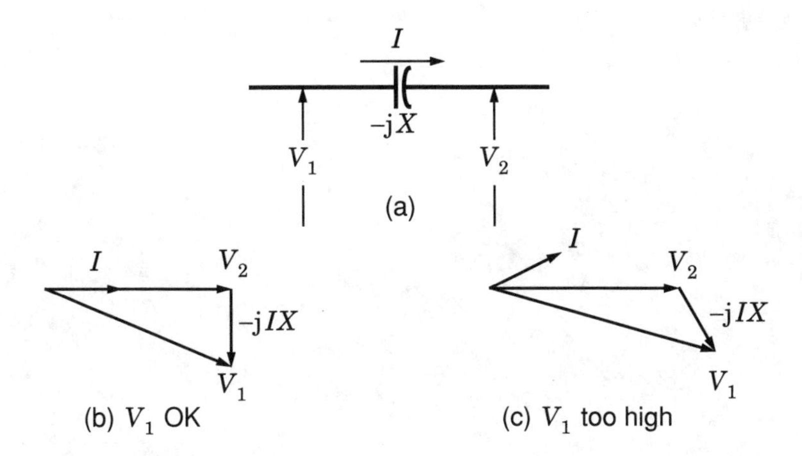

Fig. 3-6. Series capacitor voltages.

Series compensation has often been found to be cost-effective compared to shunt (SVC) compensation. Series compensation has, however, sometimes been dismissed because of unfamiliarity—and because of excessive concern over subsynchronous resonance. Subsynchronous resonance is often not a major problem, and low cost countermeasures and protective measures may be applicable.

3.3 Shunt Capacitor Banks and Shunt Reactors

Shunt capacitor banks are always bus, rather than line, connected. Shunt reactors may be either line connected or bus connected (often on the tertiary windings of autotransformers). The primary purposes of transmission system shunt compensation near load areas are voltage control and load stabilization.

Mechanically-switched shunt capacitor banks (MSCs) are installed at major substations in load areas. In this chapter, we consider applications at transmission voltages, and on tertiaries of large autotransformers. The switching is often done manually, with voltage relay backup switching when voltages are out-of-range.

Figure 3-7 shows a 523.9-kV, 342 MVAr capacitor bank at the Raver 500-kV switching station near Seattle.

For voltage stability, shunt capacitor banks are very useful in allowing nearby generators to operate near unity power factor. This maximizes fast-acting reactive reserve.

For voltage emergencies, the obvious shortcoming of shunt capacitor banks is that the reactive power output drops with the voltage squared. This is in stark contrast to the self regulating nature of series capacitors.

Fig. 3-7. 500-kV mechanically switched shunt capacitor bank. The wye-grounded bank is rated 523.9-kV, 342 MVAR, and consists of 20 parallel capacitors per group and 38 series groups. Each capacitor is 7.96-kV, 150-kVAr. *Bonneville Power Administration*.

Compared to static var compensators, mechanically switched capacitor banks have the advantage of much lower cost. Switching speeds can be quite fast. (Circuit breakers with five-cycle close, three-cycle open times are used [15].) Current limiting reactors minimize switching transients. The control, however, must minimize switching duty; this calls for generous deadbands (hysteresis) in voltage relay logic. Precise, rapid control of voltage is not possible.

Following a transmission line outage, capacitor bank energization should be delayed to allow time for line reclosing. However, capacitor switching should be before significant amounts of load are restored by transformer tap changers or distribution voltage regulators.

For the Pacific Intertie system, references 15 and 16 describe application of 500-kV switched capacitor banks to improve voltage stability. The capacitor banks are inserted by direct detection of monopolar or bipolar outages of the parallel Pacific HVDC Intertie.

References 17–20 describe recent application of voltage-controlled MSCs to improve voltage stability. Nowadays, microprocessors are preferred for the voltage control and switching logic.

There are several disadvantages to mechanically switched capacitors. For transient voltage instability, the switching may not be fast enough to prevent induction motor stalling. If voltage collapse results in a system breakup, the stable parts of the system may experience damaging overvoltages immediately following separation. Overvoltages would be aggravated by energizing of shunt capacitors during the period of voltage decay.

Generally, our discussion of shunt capacitor banks also applies to switchable shunt reactors. Because of overvoltage concerns, shunt reactors are likelier than capacitors to be applied on the tertiaries of large network autotransformers.

3.4 Static Var Systems

Static var compensators overcome the above-mentioned limitations of mechanically switched shunt capacitors and reactors. Advantages include fast, precise regulation of voltage and unrestricted, largely transient-free, capacitor bank switching. Voltage is regulated according to a slope (droop) characteristic. The slope is related to the steady state gain and is generally 1–5% over the control range. At the boost limit, the SVC becomes a shunt capacitor bank.

What are static var compensators and systems? CIGRÉ [21] distinguishes between static var compensators and static var systems. A static var system is a static var compensator with the compensator also controlling mechanical switching of shunt capacitor banks or reactors. Figure 3-8 shows a static var system schematic. The compensator may include thyristor controlled reactors (TCR), thyristor switched capacitors (TSC), and harmonic filters. The harmonic filters (for the TCR-produced harmonics) are capacitive at fundamental frequency and are 10–30% of the TCR MVAr size. A TCR, if used, is typically slightly larger than the TSC blocks so that continuous control is realized. Other possibilities are fixed capacitors (FC), and thyristor switched reactors (TSR). Usually a dedicated transformer is used with the compensator equipment at medium voltage (8–25 kV). Figure 3-9 shows the outdoor portion of a thyristor controlled reactor. Figure 3-10 shows the outdoor portion of thyristor switched capacitor.

As with series capacitors, most design aspects of static compensators are beyond our scope. We recommend references 4 and 21–25. Most discussions of power system (as opposed to industrial) applications of SVCs deal with long transmission lines and improvement in angle stability. Long lines are supported at intermediate locations by compensators which regulate the bus voltage. Each transmission segment becomes largely independent of the others, and overall steady-state angles greater than 90° are possible. Miller [4] terms this "compensation by sectioning."

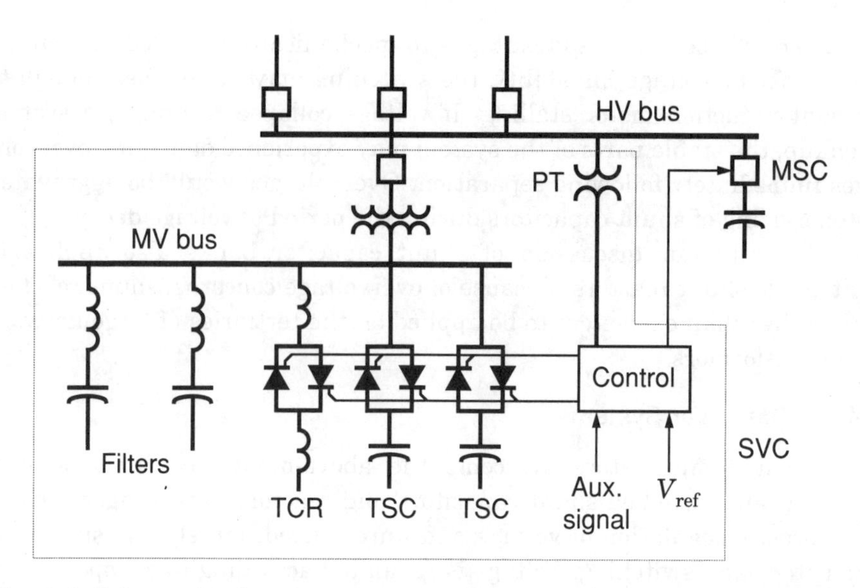

Fig. 3-8. Schematic of typical static var system.

We, however, are interested in SVCs in load areas. For preventing transient voltage instability associated with concentrations of motors, SVCs are more effective than breaker switched capacitors [4 (pages 170–172),26]. Chapter 6 provides a simulation example.

For slower forms of voltage stability, the rapid response of SVCs is not as critical—except in controlling overvoltages following breakups. Overvoltages may be due to load shedding (ac contactor release) and capacitor bank energization during the voltage decay.

The close regulation of voltage by SVCs is valuable in keeping "under the voltage." (The term "keeping under the voltage" means preventing the voltage from ever getting below normal. Low voltage means that line charging is reduced, and reactive losses are increased. Switching in a capacitor during low voltage results in less than full capacitor output. Recovering to normal voltage without shedding load may be difficult.)

During normal conditions, SVCs should be operated with inductive or floating output so that rapid capacitive boost is available for disturbances. (Considering losses, floating output favors a TCR–TSC compensator over a TCR–FC compensator.) To achieve capacitive reserve, the SVC can order mechanical switching of nearby shunt capacitor banks and shunt reactors.

A reactive power or susceptance regulator may also be used to maintain the desired output during steady-state conditions, allowing voltage to vary with a deadband of several percent. Figure 3-11 shows steady-state and dynamic voltage versus reactive power characteristics. Following a dis-

Fig. 3-9. Outdoor portion of thyristor controlled reactor rated 163 MVAr. For each phase, half-sized reactors are on each side of the indoor thyristor valves. The TCR is delta connected. *ABB*.

Fig. 3-10. Outdoor portion of thyristor switched capacitor rated 121 MVAr. The TSCs are delta connected and include current limiting reactors. *ABB*.

turbance, SVCs will respond faster than other voltage regulating controls. To reposition for the next disturbance, and to coordinate with slower equipment, the SVC reactive power output is slowly returned to its setpoint (Figure 3-12); this allows generator excitation control, voltage-controlled mechanically switched capacitors and reactors, and network LTC transformers to respond. Generator excitation control can include slow control of high-side voltage. Time constants of reactive power regulators are tens of seconds or minutes.

The SVC slope setting is important in coordination with other voltage control equipment, especially in the absence of a reactive power regulator. Following a disturbance, a large slope setting reduces SVC response, causing a larger voltage drop. The reduced SVC response may allow voltage-controlled shunt capacitor banks to switch on.

The need for static compensators in load areas relates to the amount of local generation. As load grows, reactive power reserve of generators may become so small that other fast acting reactive reserve becomes desirable.

Generators and static compensators are valuable *pilots* which can warn of impending voltage collapse. Activation of reactive power reserve and closeness to boost limits can be monitored, and are sensitive indicators of voltage security problems.

Example 3-5 [27]. Figure 3-13 shows, conceptually, V–Q characteristics for both constant and voltage sensitive loads. The initial post-disturbance system characteristic is for voltage sensitive loads. The final post-disturbance characteristic is after tap changing has restored load to pre-disturbance levels. Immediately following the disturbance, the operating point is at a very low voltage. Since the final constant power characteristic does not intersect the $Q = 0$ axis, the load restoration by tap changing causes collapse.

Voltage stability can be maintained by a static var system. Figure 3-14 shows SVC and mechanically switched capacitor characteristics superimposed on the system V–Q characteristics. The SVC characteristic shown on Figure 3-11 is simply rotated 90° clockwise. The SVC slope can be set to achieve a desired minimum voltage (V_{min}). Some reactive reserve is required. Before load restoration, a MSC must be energized to bias the SVC to intersect with the constant power characteristic—again with reserve margin. The mechanical switching can be directed by the static compensator controls.

SVC Modeling. References 21, 28, and 29 describe models for power flow and dynamic simulations. Reference 30 provides a good description of controls and models for a recent installation.

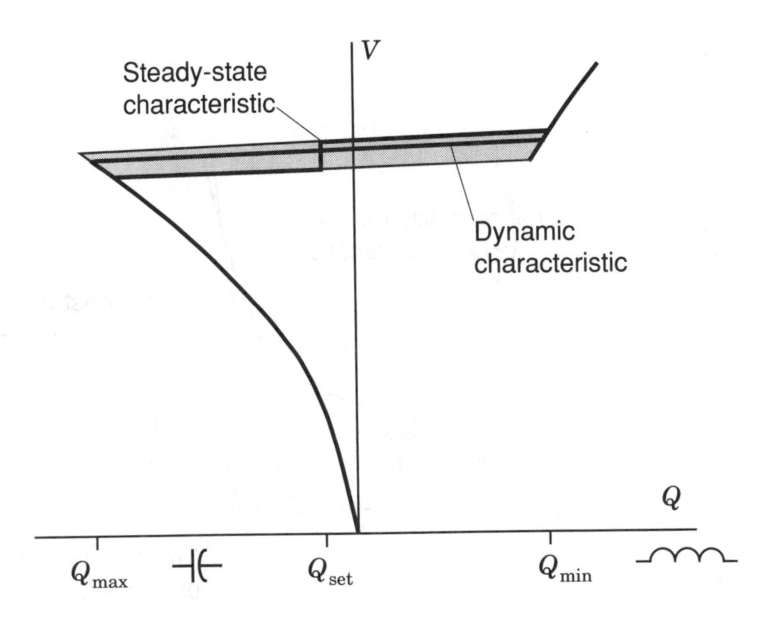

Fig. 3-11. SVC high-voltage side voltage/reactive power characteristic with reactive power or susceptance regulator.

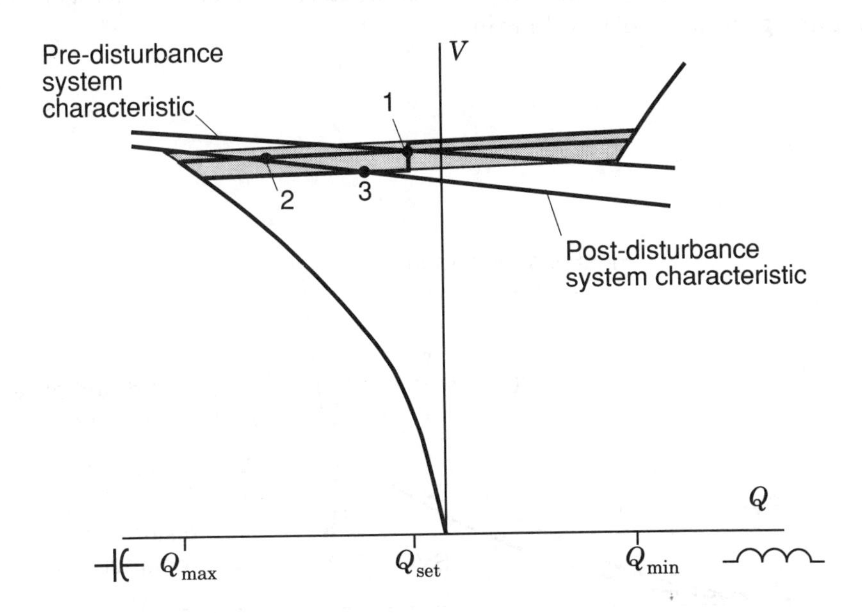

Fig. 3-12. SVC reactive power control. The SVC moves along dynamic characteristic from point 1 to point 2 immediately following a disturbance. Reactive power output then slowly moves to point 3 to reestablish reactive reserve and allow other controls to raise the system characteristic.

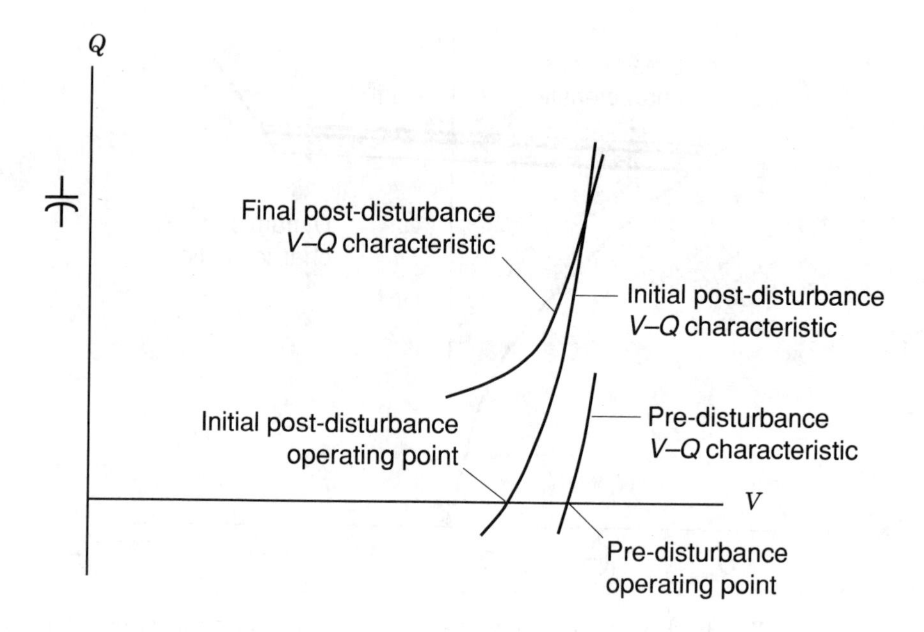

Fig. 3-13. Pre- and post-disturbance V–Q characteristics. At the lower voltages, the final post-disturbance V–Q characteristic has tap changers at limit and generator current limiting.

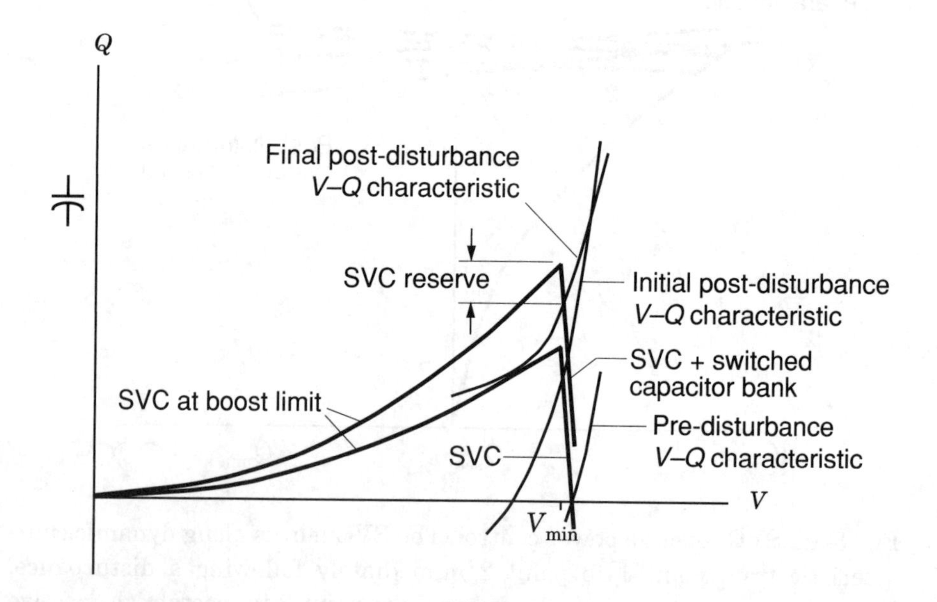

Fig. 3-14. Static var system (SVC and MSC) for stabilization.

For specialized study of voltage control, such as for coordination of reactive power compensation, power flow simulation should include slope or droop representation. The slope can be modeled by connecting the SVC to an auxiliary bus separated from the SVC high voltage bus by a reactance equal to the per unit slope [21]. The auxiliary bus is a PV bus controlling the voltage at the SVC high voltage bus. The SVC should become a capacitor at its boost limit; with older power flow programs this can be achieved by representing a thyristor controlled reactor/fixed capacitor type of SVC.

If a reactive power or susceptance regulator is used, the SVC model in a power flow simulation must correspond to the point in time that is simulated.

For dynamic simulation (small-signal analysis, transient stability, longer-term dynamics) relatively simple models usually suffice. Figure 3-15 shows the model structure for a typical SVC.

3.5 Comparisons between Series and Shunt Compensation

Let's compare series and shunt compensation.

Some series compensation advantages are:

- Series capacitors are inherently self regulating; a control system is not needed.
- For equivalent performance, series capacitors are often less costly than SVCs; losses are very low.
- For voltage stability, series capacitors lower the critical or collapse voltage.
- Series capacitors have significant time-overload capability.
- Series capacitors and switched series capacitors can be used to control loading of parallel lines to minimize active and reactive losses.

Some series compensation disadvantages are:

- To reinforce an established grid, many parallel lines may have to be series compensated.
- Series capacitors are line connected; compensation is removed for outages, and capacitors in parallel lines may be overloaded.
- During heavy loading (outages of parallel lines), the voltage on one side of the series capacitor may be out-of-range.
- Shunt reactors may be needed for light load compensation.
- Subsynchronous resonance may call for expensive countermeasures.

Some static var compensator advantages are:

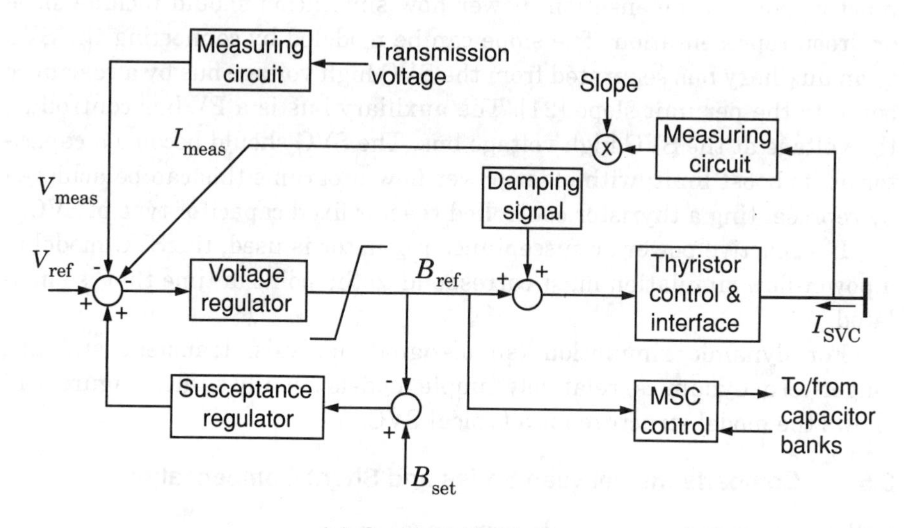

(a) General structure

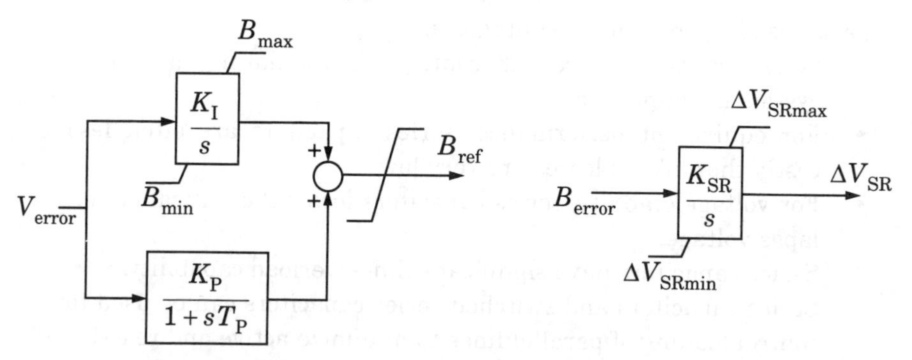

(b) Proportional–integral voltage regulator (c) Susceptance regulator

Fig. 3-15. Static var system model. The voltage regulator is often purely integral control. The current may be obtained by multiplication of B_{ref} and V_{meas}. The susceptance regulator limits determine the voltage deadband. Other signals may be active during disturbances to improve transient stability and transient damping.

- SVCs provide direct control of voltage; this is very valuable when there is little generation in the load area.
- The remaining capacitive capability of a SVC is a good indication of proximity to voltage instability.
- SVCs provide rapid control of temporary overvoltages.

Some static var compensator disadvantages are:

- SVCs have limited overload capability; a SVC is a capacitor bank at its boost limit.
- The critical or collapse voltage becomes the SVC-regulated voltage; instability usually occurs once an SVC reaches its boost limit.
- SVCs are expensive; shunt capacitor banks should first be used to allow unity power factor operation of nearby generators.

The best design in some cases may be a combination of series and shunt compensation.

In the early 1960s, E. C. Starr compared series and shunt compensation. Figures 3-16 and 3-17 show some inherent properties on P–V curves. For switchable shunt compensation (nowadays achievable with thyristor switched capacitors), the critical voltage increases with more compensation. Series compensation has the opposite property.

We can provide some idea of present day compensation equipment costs. The data given below are to furnish and install equipment at an existing substation. Overhead costs are excluded.

500-kV series capacitors:	\$2.8 million + \$7.50/KVAr
500-kV shunt reactors:	\$18.3/kVAr
500-kV shunt capacitors:	\$1.3 million + \$4.10/kVAr
230-kV shunt capacitors:	\$0.28 million + \$4.10/kVAr
Large SVC:	\$30–50/kVAr

The fixed cost of the shunt capacitors is circuit breaker cost. The range of SVC costs is related to specific configuration and the complexity of the project. All costs are subject to variation because of monetary exchange rates, level of competition, special requirements, and other factors.

3.6 Synchronous Condensers

Because of higher initial and operating costs, synchronous condensers are generally not competitive with static var compensators. The capital cost may be 20–30% higher than SVCs. The full-load losses of condensers (and generators) are around 1.5%, and the no-load losses (floating output) are around 0.5%.

Technically, synchronous condensers have some advantages over SVCs in voltage-weak networks. Following a drop in network voltage, the increase in condenser reactive power output is instantaneous (same for generator reactive output). The subsequent decay of internal voltage or flux (armature reaction) is countered by excitation control. As discussed in

Chapter 5, condensers and generators have tens of seconds of overload capability. In contrast to the voltage squared capacitor characteristic of SVCs at full boost, condensers can maintain rated current at full boost.

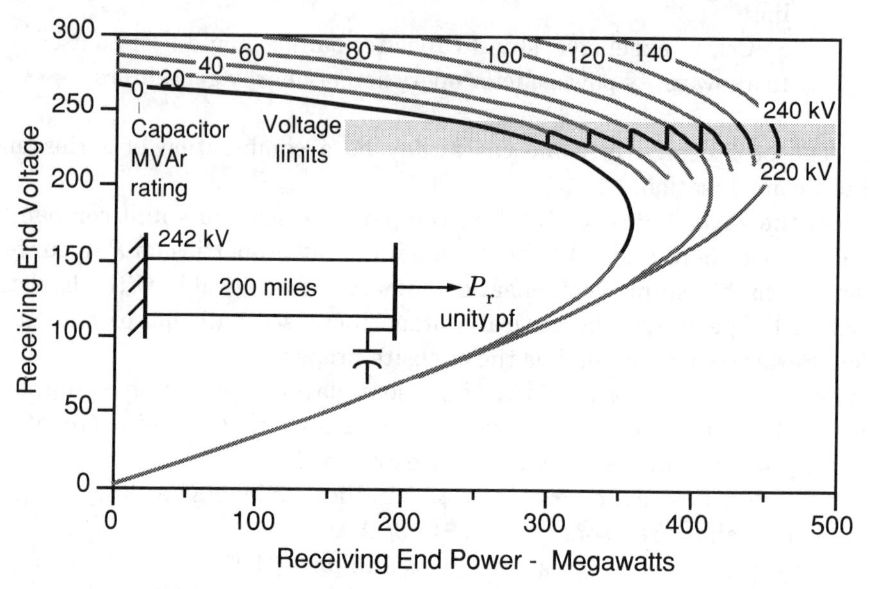

Fig. 3-16. Characteristics of shunt compensation.

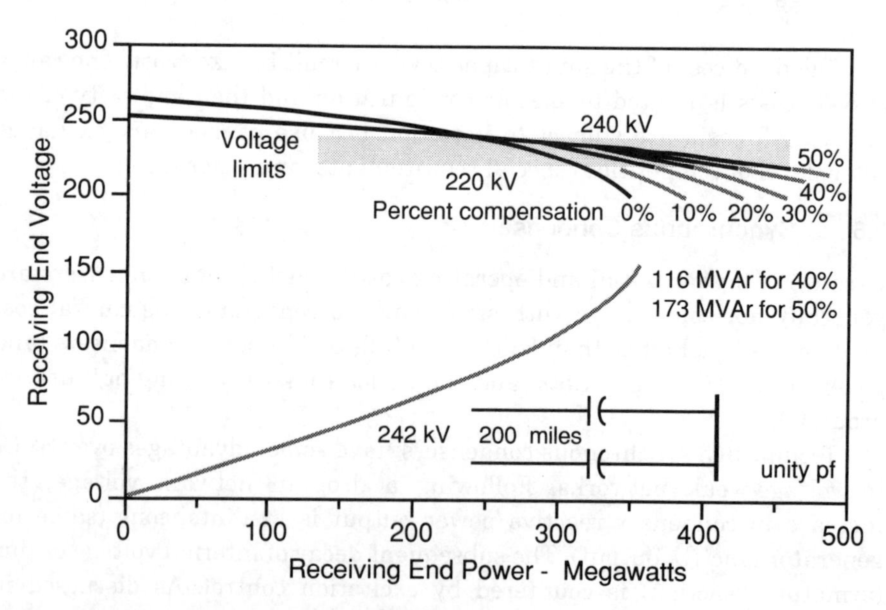

Fig. 3-17. Characteristics of series compensation.

For voltage stability improvement, Tokyo Electric Power Company has chosen synchronous condensers over static var compensators. One reason is the lower critical voltage at the maximum power point [31].

As a solution to voltage stability problems, generators (especially combustion turbines) have been declutched to run as synchronous condensers [17,32–34].

Most synchronous condenser applications are now associated with high voltage direct current (HVDC) installations. They are sometimes necessary to increase the network strength (short circuit capacity) and to improve (or provide) commutation voltage. (To date, all HVDC systems use line-commutated inverters.) Manitoba Hydro recently installed three +300/-165 MVAr synchronous condensers at the Nelson River project inverter station [35]. The machines have 360 MVAr rating for temperatures below 20°C.

The companion book [28] describes modeling of synchronous machines.

3.7 Transmission Network LTC Transformers

Transmission network autotransformers with load tap changers provide voltage and reactive power control. LTC transformers can be either manually or automatically controlled. Most North American LTC autotransformers are manually controlled. (Many North American utilities purchase large autotransformers with fixed or no-load taps.)

Tap changers regulate the low voltage side. For example, automatic tap changing on 500/230-kV autotransformers would regulate the 230-kV voltage. This supports capacitor banks and line charging on the lower voltage network, and reduces reactive power losses on the lower voltage network. Transmission system transformer tap changing time delays should be short compared to the time delay for tap changing for bulk power delivery transformers and distribution voltage regulators. This will result in faster load restoration compared to fixed tap transformers or manually controlled LTC transformers.

The EHV-side voltages will sag because of the tap changing, resulting in higher EHV network reactive losses. To help prevent this, tap changing should be coordinated with switching of transmission network shunt capacitor banks and shunt reactors. One method is to switch shunt reactive devices rather than transformer taps if the disturbance lowers voltage on both sides of the LTC autotransformer [19].

Tap changing is further discussed in the next chapter in conjunction with subtransmission and distribution networks.

We have introduced aspects of transmission :tive power compensation and control pertaining directly to ·ility. Later chapters will build on this introduction.

References

1. H. P. St. Clair, "Practical Concepts in Capability and of Transmission Lines," *AIEE Transactions, Part III Power Apparems*, Vol. 72, pp. 1152–1157, December 1953.
2. R. D. Dunlop, R. Gutman, and P. P. Marchenko, "Aelopment of Loadability Characteristics for EHV and UHV Tr.ines," *IEEE Transactions on Power Apparatus and Systems*, No. 2, pp. 606–617, March/April 1979.
3. H. W. Dommel, *Notes on Power System Analysis*, Unitish Columbia, 1975.
4. T. J. E. Miller, editor, *Reactive Power Control in Elect*ohn Wiley & Sons, New York, 1982.
5. G. D. Breuer, H. M. Rustebakke, R. A. Gibley, and H. Jr., "The Use of Series Capacitors to Obtain Maximum EHV Transbility," *IEEE Transactions on Power Apparatus and Systems*, No. 11, pp. 1090–1101, November 1964.
6. F. Iliceto, E. Cinieri, M. Cazzani, and G. SantagostinVoltages and Currents in Series-Compensated EHV Lines," *Proceel*. 123. No. 8, pp. 811–817, August 1976.
7. F. Iliceto and E. Cinieri, "Comparative Analysis of Set Compensation Schemes for AC Transmission Systems," *IEEs on Power Apparatus and Systems*, Vol. PAS-96, No. 6, pp. 181nber/December 1977.
8. E. W. Kimbark, "Improvement of System Stability bries Capacitors," *IEEE Transactions on Power Apparatus and S*AS-85, No. 2, pp. 180–188, February 1966.
9. R. Allustiarti, H. Hoexter, P. Lai, J. Samuelsson,, and R. G. Rocamora, "Design and Operating Performance of l-Oxide-Protected Series Capacitor Banks on the Table Mountain*IEEE Transactions on Power Delivery*, Vol. 3, No. 4, pp. 1951–1958.
10. J. M. Barcus, S. A. Miske, Jr., A. P. Vitols, H. M. MayG. Peterson, "The Varistor Protected Series Capacitors at the 5iew Substation," *IEEE Transactions on Power Delivery*, Vol. 3, N–1985, October 1988.
11. G. E. Lee and D. L. Goldsworthy, "Equipment and · BPA's New Intertie Series Capacitors," *Proceedings of the Ameriaference*, Vol. 54-I, pp. 664–670, 1992.
12. IEEE Committee Report, *Analysis and Control of Ss Resonance*, IEEE 76 CH1066-0-PWR, 1976.
13. IEEE Committee Report, *Symposium on Countermeasynchronous Resonance*, IEEE 81TH0086-9-PWR, 1981.
14. P. M. Anderson, B. L. Agrawal, and J. E. Van Ness, *Ss Resonance in Power Systems*, IEEE Press, 1990.

15. B. C. Furumasu and R. M. Hasibar, "Design and Installation of 500-kV Back-to-Back Shunt Capacitor Banks," *IEEE Transactions on Power Delivery*, Vol. 7, No. 2, pp. 539–545, April 1992.
16. W. Mittelstadt, C. Taylor, M. Klinger, J. Luini, J. McCalley, and J. Mechenbier, "Voltage Instability Modeling and Solutions as Applied to the Pacific Intertie," *CIGRÉ*, paper 38-230, 1990.
17. W. B. Jervis, J. G. P. Scott, and H. Griffiths, "Future Application of Reactive Compensation Plant on the CEGB System to Improve Transmission Network Capability," *Proceedings of 33rd CIGRÉ Session*, Vol. II, paper 38-06, 1988.
18. I. M. Welch, P. H. Buxton, and D. S. Crisford, "Integration of Reactive Power Compensation Equipment within Substations," *CIGRÉ*, paper 23-202, 1990.
19. G. N. Allen, V. E. Henner, and C. T. Popple, "Optimization of Static Var Compensators and Switched Shunt Capacitors in a Long Distance Interconnection," *Proceedings of 33rd CIGRÉ Session*, Vol. II, paper 38-07, 1988.
20. S. Koishikawa, S. Ohsaka, M. Suzuki, T. Michigami, and M. Akimoto, "Adaptive Control of Reactive Power Supply Enhancing Voltage Stability of a Bulk Power Transmission System and a New Scheme of Monitor on Voltage Security," *CIGRÉ*, paper 38/39-01, 1990.
21. CIGRÉ WG 38-01, *Static Var Compensators*, CIGRÉ, Paris, 1986.
22. K. Hanson, *Application Guide for Static Var Compensators*, Electricity Supply Board, Dublin, Ireland, 1985.
23. Canadian Electrical Association, *Static Compensators for Reactive Power Control*, Cantext Publications, Winnipeg, 1984.
24. IEEE Committee Report, *Application of Static Var Systems for System Dynamic Performance*, IEEE 87TH0187-5-PWR, 1987.
25. D. L. Osborn, "Factors for Planning a Static VAR System," *Electric Power Systems Research*, 17 (1989), pages 5–12.
26. A. E. Hammad and M. Z. El-Sadek, "Prevention of Transient Voltage Instabilities due to Induction Motor Loads by Static VAR Compensators," *IEEE Transactions on Power Systems*, Vol. 4, No. 3, pp. 1182–1190, August 1989.
27. CIGRÉ Working Group 38.01, "Planning Against Voltage Collapse," *Electra*, pp. 55–75, March 1987.
28. P. Kundur, *Power System Stability and Control*, McGraw-Hill, 1993.
29. IEEE Special Stability Controls Working Group, "Static Var Compensator Models for Power Flow and Dynamic Performance Simulation," paper 93 WM 173-5 PWRS, IEEE/PES 1993 winter meeting.
30. D. Dickmander, B. Thorvaldsson, G. Strömberg, D. Osborn, A. Poitras, and D. Fisher, "Control System Design and Performance Verification for the Chester, Maine Static Var Compensator," *IEEE Transactions on Power Delivery*, Vol. 7, No. 3, pp. 1492–1503, July 1992.
31. Y. Sekine, K. Takahashi, Y. Ichida, Y. Ohura, and N. Tsuchimori, "Method of Analysis and Assessment on Power System Voltage Phenomena, and Improvements Including Control Strategies for Greater Voltage Stability Margins," *CIGRÉ*, paper 38-206, 1992.
32. F. Iliceto, E. Einieri, F. Gatta, and A. Erkan, "Optimal Use of Reactive Power Resources for Voltage Control in Long Distance EHV Transmission: Applications to the Turkish 420-kV System," *Proceedings of 33rd CIGRÉ Session*, Vol. II, paper 38-03, 1988.

33. M. G. Dwek, Study Group 38 discussion, *Proceedings of 33rd CIGRÉ Session*, Vol. II, 1988.

34. M. J. Lefrancois and J. C. McGough, "Highlights of the B.C. Hydro Burrard Generating Units Operating as Synchronous Condensers," *Proceedings of the American Power Conference*, April 1990.

35. C. V. Thio and J. B. Davies, "New Synchronous Compensators for the Nelson River HVDC System—Planning Requirements and Specification," *IEEE Transactions on Power Delivery*, Vol. 6, No. 2, pp. 922–928, April 1991.

4

Power System Loads

God is in the Details
Ludwig Mies Van Der Rohe

Voltage stability depends on the details—particularly the load characteristics. In our context, loads include the subtransmission and distribution networks that connect consumers to the transmission networks and generation. To be expert in voltage stability analysis, we must understand the load characteristics and be able to model them. We must also understand aspects of distribution system engineering.

4.1 Overview of Subtransmission and Distribution Networks

Subtransmission networks usually operate at voltages such as 69-kV, 115-kV, and 138-kV that are no longer considered "main-grid" and that may have many taps for load delivery. Large industrial customers are often served directly from the subtransmission (or even the transmission) networks. In some systems, the subtransmission may be radial from the main-grid, and in others there may be loop back to main-grid substations. The characteristics of subtransmission networks are not much different from the transmission networks described in Chapter 3. In this chapter we concentrate on distribution networks.

A good reference on distribution system engineering is the book by Gönen [1].

Figure 4-1 shows a one-line diagram of a typical radial distribution network. The power is delivered to the distribution network at bulk power delivery substations. The primary voltage is typically 115-kV, 138-kV, or 230-kV. The secondary or distribution voltage is 4 to 35 kV, with 12 kV and higher voltages becoming dominant.

Transformers. The bulk power delivery transformers may have load tap changers (LTC transformers). Or there may be a voltage regulator in series with the transformer on the secondary side.

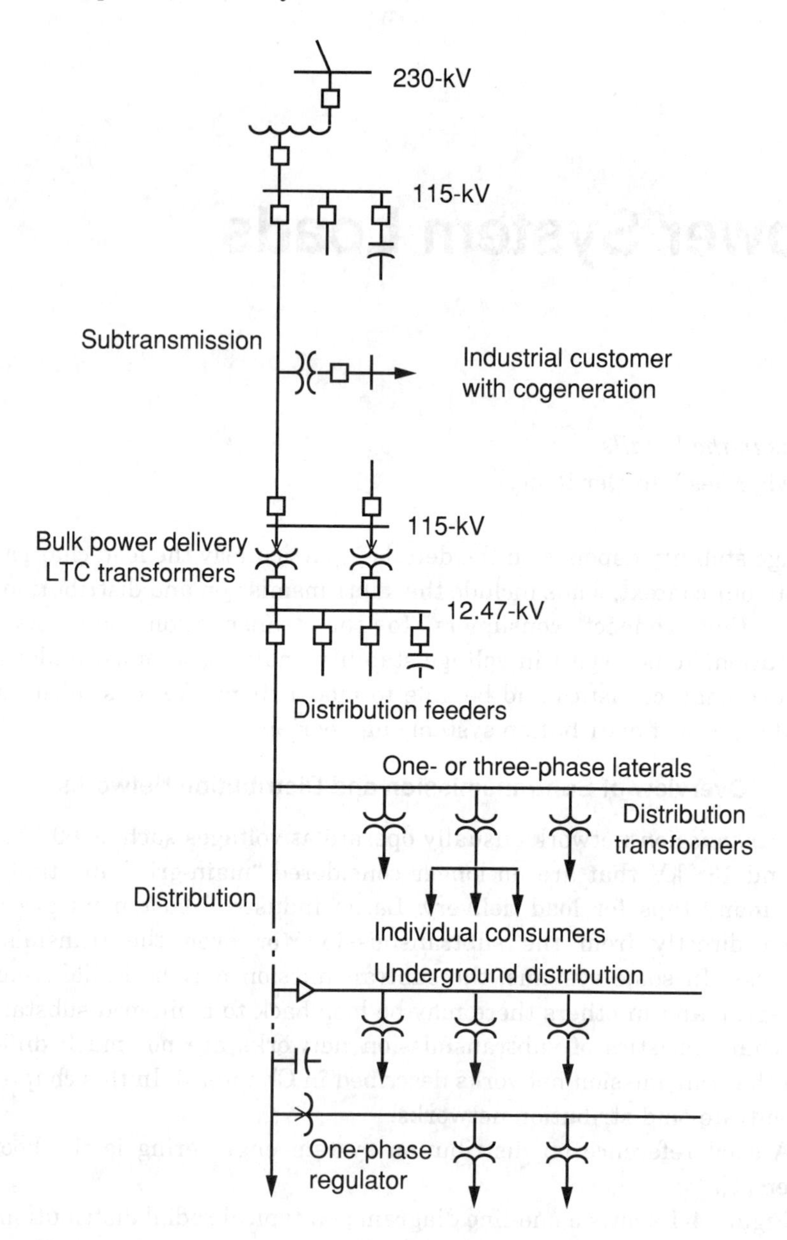

Fig. 4-1. One-line diagram of a typical radial distribution system. Not shown are fuses, reclosers, and switched capacitors.

For voltage stability, the impedance of transformers is important. Impedances (leakage reactances) of bulk power delivery transformers are about 8–11% on the transformer base. Impedances of distribution transformers are 2–4%. References 1 and 2 provide tables of transformer impedances. Transformers, particularly the bulk power delivery transformers, represent a large share of the distribution system impedance.

Many distribution transformers operate with some degree of saturation (several percent exciting current at rated voltage). Exciting current is reduced during voltage sags, accounting for the high reactive power voltage sensitivity (3–6% reduction in reactive power for a 1% reduction in voltage) often observed in field tests.

Feeder characteristics. Distribution feeder circuits leave the bulk power delivery substation. Where feeders are of different length, the transformer tap changing regulator is often replaced with individual feeder tap changing voltage regulators (essentially autotransformers). Line-drop compensation, described in a later section is often used.

Positive sequence X/R ratios are much smaller for overhead distribution circuits than transmission circuits. X/R ratios less than unity are common, particularly on older circuits and on laterals. A rule of thumb (related to asymmetrical fault currents) is that X/R ratios should not exceed five [1]. Equations 1.1 through 1.4 no longer apply, and active and reactive power relations are more complex. The differences can be easily seen from voltage phasor diagrams (Figure 4-2).

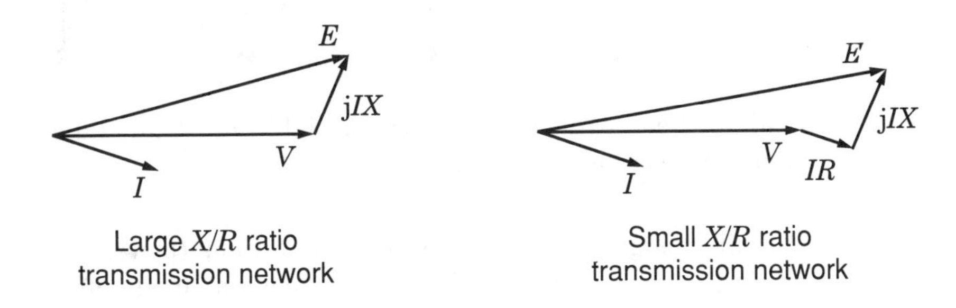

Fig. 4-2. Phasor diagrams for high and low X/R ratios.

Long feeders, which generally serve residential or farm loads, may require additional voltage regulators, possibly with line-drop compensation. Fixed and switched shunt capacitors can also be installed along the feeder to control voltage. If the loads are uniformly distributed along the feeder, fixed capacitors are about two-thirds of the distance to the end of the feeder [1]. Switched capacitors may be close to the end of the feeder. Switched capacitors are now less expensive than regulators and have lower losses. They are increasingly used.

Underground distribution, including underground residential distribution (URD), is widely used. The cables have much lower reactance, higher X/R ratios, and higher capacitance.

EPRI research on load modeling surveyed forty utilities regarding load composition and distribution feeder characteristics, including feeder lengths and impedances [2]. Based on data from three Texas utilities, the researchers developed the typical overhead residential feeder shown on Figure 4-3. Feeders for commercial and industrial loads are generally shorter.

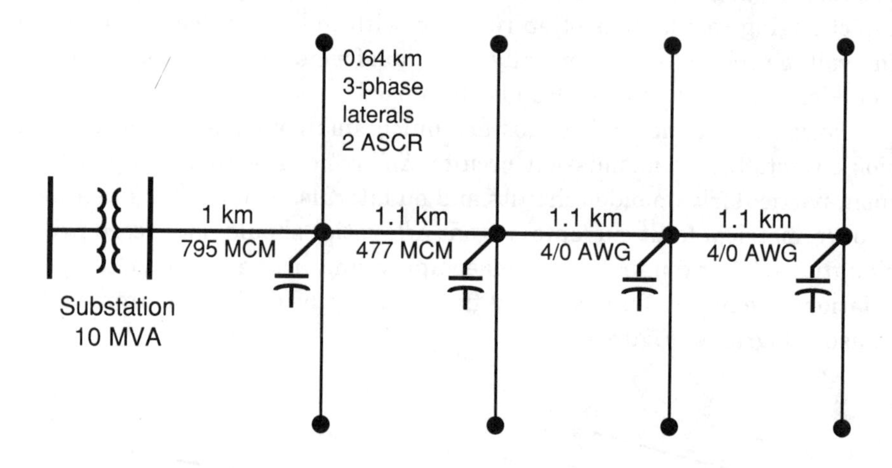

Fig. 4-3. Typical overhead residential feeder [2]. Light industrial and commercial load may be served along trunk.

For the circuit of Figure 4-3, sensitivity computations (power versus voltage for static load models without regulators) evaluated the effect of feeder impedance changes. Results were insensitive to lateral impedance changes. For winter load compositions, results were also insensitive to changes in trunk impedance (winter loads are highly voltage sensitive). For southern U.S. residential summer load compositions with 77% air conditioning and refrigeration load (compressor load), results were highly sensitive to trunk impedance changes—particularly for reactive power. Figure 4-4 shows the results. The increase in reactive power with voltage sag is obviously detrimental to voltage stability and is a characteristic of motors driving compressor loads. Compressor loads require nearly constant torque regardless of motor speed.

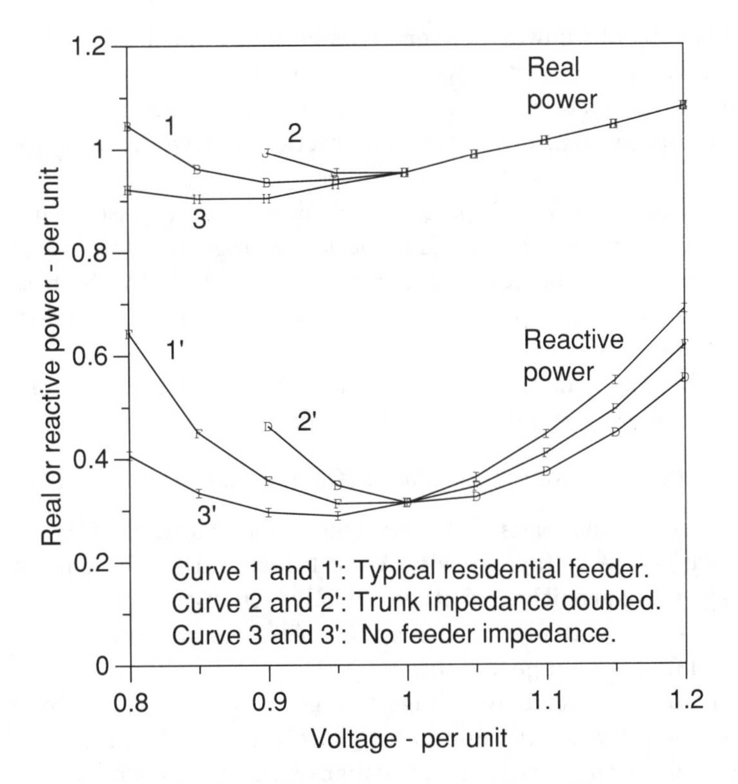

Fig. 4-4. Effect of feeder impedance on active and reactive power responses to substation voltage changes [2, Volume 3]. Residential feeder (Fig. 4-3) with large central air conditioning load. One per unit power equals 10 MVA.

Other sensitivity cases showed that linearly increasing load (triangular-shaped feeder configuration) is more severe than the uniformly distributed load (rectangular-shaped feeder configuration) of Figure 4-4.

Example 4-1. For a 12.47-kV, 10 MVA residential feeder, compare per unit feeder impedance with the impedance of the transformers.

Solution: Base impedance for 12.47-kV and 10 MVA is $12.47^2/10 = 15.55$ ohms. Typical feeder impedance may be about 0.5 Ω/km or 0.03 pu/km. The effective impedance of a 5 km feeder with uniformly distributed load is about $0.03 \times 5/2 = 0.075$ per unit. This compares with about 0.1 per unit and 0.03 per unit impedances for the bulk power delivery and distribution transformers, respectively. Total transformer and feeder impedance for residential feeders typically may be around 0.2 per unit.

Summary. For voltage stability, we are especially interested in the characteristics of longer feeders (residential and rural) that add significant impedance between bulk power delivery transformers and loads. We have not discussed distribution networks for more concentrated loads.

Distribution system engineering practices vary considerably from company to company. Within a company, practices evolve over time, resulting in feeders having different characteristics.

Several practices are important for voltage stability. Some utilities do not use LTC transformers or distribution voltage regulators, relying on voltage controlled capacitors. Other utilities use LTCs on both transmission/subtransmission transformers and subtransmission/distribution transformers.

Later in this chapter, we describe distribution reactive power compensation and voltage regulation in more detail.

4.2 Static and Dynamic Characteristics of Load Components

We now describe the loads and their aggregated characteristics. Of critical importance is the voltage sensitivity of loads. (Although some loads are also frequency sensitive, voltage usually changes much more than frequency—even with electrical islanding. Also, frequency sensitivity is not directly related to voltage stability.)

The response of loads to voltage changes occurring over many minutes can affect voltage stability. For transient voltage stability, and for the final stages of a slower occurring voltage collapse, the dynamic characteristics of loads such as induction motors are critical.

For the slower forms of voltage instability, a key question is whether or not—in power flow simulations—the normal constant power (voltage independent) load model is valid. In some systems there are enough control devices to keep loads constant, and the load characteristics are then unimportant until the controls reach limits. In large-scale simulation, constant power loads are usually represented on the high voltage side of bulk power delivery substations; the neglect of transformer and feeder impedances may compensate for the conservative constant power assumption. Constant power may be reasonable for approximate static analysis when a significant proportion of the load is motors (e.g., summertime with air conditioning). The alternative to constant power load models is representation of voltage sensitive loads plus regulators (i.e., tap changers) which restore load.

LOADSYN program. EPRI has funded several load modeling projects, the latest resulting in the LOAD SYNthesis (LOADSYN) computer program [3–5]. Table 4-1 summarizes parameters for models of many load components. The voltage and frequency sensitivity for static models are in exponential form:

Table 4-1 - check the errata sheet

			Static characteristics									Dynamic Characteristics								
Component	PF	Pv	Pf	Qv	Qf	N_m	PF_{nm}	Pv_{nm}	Pf_{nm}	Qv_{nm}	Qf_{nm}	R_s	X_s	X_m	R_r	X_r	A	B	H	LF_m
Resistance space heater	1.0	2.0	0.0	0.0	0.0	0.0	-	-	-	-	-	-	-	-	-	-	-	-	-	-
Heat pump space heating	0.84	0.2	0.9	2.5	-1.3	0.9	1.0	2.0	0.	0.	0.	.33	.076	2.4	.048	.062	0.2	0.	0.28	0.6
Heat pump central air cond.	0.81	0.2	0.9	2.5	-2.7	1.0	-	-	-	-	-	.33	.076	2.4	.048	.062	0.2	0.	0.28	0.6
Central air conditioner	0.81	0.2	0.9	2.2	-2.7	1.0	-	-	-	-	-	.33	.076	2.4	.048	.062	0.2	0.	0.28	0.6
Room air conditioner	0.75	0.5	0.6	2.5	-2.8	1.0	-	-	-	-	-	.10	.10	1.8	.09	.06	0.2	0.	0.28	0.6
Water heater	1.0	2.0	0.0	0.0	0.0	0.0	-	-	-	-	-	-	-	-	-	-	-	-	-	-
Range	1.0	2.0	0.0	0.0	0.0	0.0	-	-	-	-	-	-	-	-	-	-	-	-	-	-
Refrigerator and freezer	0.84	0.8	0.5	2.5	-1.4	0.8	1.0	2.0	0.	0.	0.	.056	.087	2.4	.053	.082	0.2	0.	0.28	0.5
Dishwasher	0.99	1.8	0.0	3.5	-1.4	0.8	1.0	2.0	0.	0.	0.	.11	.14	2.8	.11	.065	1.0	0.	0.28	0.5
Clothes washer	0.65	0.08	2.9	1.6	1.8	1.0	-	-	-	-	-	.11	.12	2.0	.11	.13	1.0	0.	0.69	0.4
Incandescent lighting	1.0	1.54	0.0	0.0	0.0	0.0	-	-	-	-	-	-	-	-	-	-	-	-	-	-
Clothes dryer	0.99	2.0	0.0	3.3	-2.6	0.2	1.0	2.0	0.	0.	0.	.12	.15	1.9	.13	.14	1.0	0.	0.11	0.4
Colored television	0.77	2.0	0.0	5.2	-4.6	0.0	-	-	-	-	-	-	-	-	-	-	-	-	-	-
Furnace fan	0.73	0.08	2.9	1.6	1.8	1.0	-	-	-	-	-	-	-	-	-	-	-	-	-	-
Commercial heat pump	0.84	0.1	1.0	2.5	-1.3	0.9	1.0	2.0	0.	0.	0.	.53	.83	1.9	.036	.068	0.2	0.	0.28	0.6
Heat pump commercial A/C	0.81	0.1	1.0	2.5	-1.3	1.0	-	-	-	-	-	.53	.83	1.9	.036	.068	0.2	0.	0.28	0.6
Commercial central A/C	0.75	0.1	1.0	2.5	-1.3	1.0	-	-	-	-	-	.53	.83	1.9	.036	.068	0.2	0.	0.28	0.6
Commercial room A/C	0.75	0.5	0.6	2.5	-2.8	1.0	-	-	-	-	-	.10	.10	1.8	.09	.06	0.2	0.	0.28	0.6
Fluorescent lighting	0.90	1.0	1.0	3.0	-2.8	0.0	-	-	-	-	-	-	-	-	-	-	-	-	-	-
Pumps, fans, other motors	0.87	0.08	2.9	1.6	1.8	1.0	-	-	-	-	-	.079	.12	3.2	.052	.12	1.0	0.	0.7	0.7
Electrolysis	0.90	1.8	-0.3	2.2	0.6	0.0	-	-	-	-	-	-	-	-	-	-	-	-	-	-
Arc furnace	0.72	2.3	-1.0	1.61	-1.0	0.0	-	-	-	-	-	-	-	-	-	-	-	-	-	-
Small industrial motors	0.83	0.1	2.9	0.6	-1.8	1.0	-	-	-	-	-	.031	.10	3.2	.018	.18	1.0	0.	0.7	0.6
Large industrial motors	0.89	0.05	1.9	0.5	1.2	1.0	-	-	-	-	-	.013	.067	3.8	.009	.17	1.0	0.	1.5	0.8
Agricultural water pumps	0.85	1.4	5.6	1.4	4.2	1.0	-	-	-	-	-	.025	.088	3.2	.016	.17	1.0	0.	0.8	0.7
Power plant auxiliaries	0.80	0.08	2.9	1.6	1.8	1.0	-	-	-	-	-	.013	.14	2.4	.009	.12	1.0	0.	1.5	0.7

$$P = P_0 \left[\frac{V}{V_0}\right]^{Pv} \left[\frac{f}{f_0}\right]^{Pf} \tag{4.1}$$

$$Q = Q_0 \left[\frac{V}{V_0}\right]^{Qv} \left[\frac{f}{f_0}\right]^{Qf} \tag{4.2}$$

These static models may be valid for only a limited voltage range (say, ±10%). For motors and discharge lighting, the models are inadequate for large voltage deviations. Representation of loads by exponential models with exponent values less than 1.0 in a dynamic simulation is questionable [6].

First, the load power factor is listed, followed by the static voltage and frequency dependency exponents. N_m is the part of the load that is motor. Next is listed the power factor and static model parameters for the non-motor part of the load. For example, a heat pump is 90% motor and 10% resistive. Finally, parameters for the dynamic motor model are listed. The motor load factor (LF_m) is the ratio of motor MW load to motor MVA rating. This is an important, but often overlooked, parameter. LF_m is an average value for cyclic load components such as clothes washers and dryers. The *A* and *B* parameters describe the mechanical load torque characteristics as described by Equation 4.8.

Figure 4-5 shows the induction motor steady-state equivalent circuit corresponding to the data of Table 4-1.

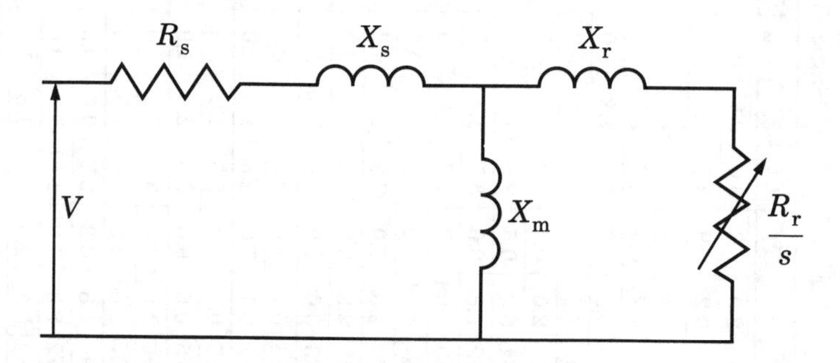

Fig. 4-5. Induction motor steady-state equivalent circuit corresponding to data of Table 4-1.

A third order model is usually adequate for aggregated motors in bulk system dynamic simulation. Stator transients are neglected and electrical transients in one rotor circuit per axis are represented. Figure 4-6 shows the transient state equivalent circuit consisting of a voltage behind transient impedance. The equations for the induction motor model are [7,8]:

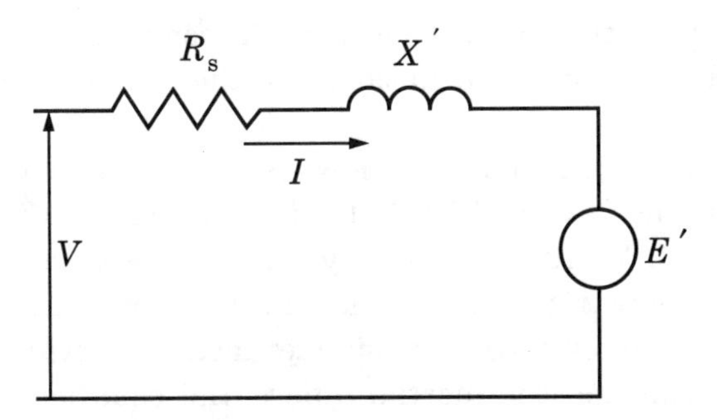

Fig. 4-6. Induction motor transient-state equivalent circuit.

$$\frac{dE_q'}{dt} = -\omega_o s E_d' - \frac{1}{T_o'} E_q' + \frac{X - X'}{T_o'} i_d \tag{4.3}$$

$$\frac{dE_d'}{dt} = -\frac{1}{T_o'} E_d' + \omega_o s E_q' - \frac{X - X'}{T_o'} i_q \tag{4.4}$$

$$i_d = \frac{1}{R_s^2 + X'^2} [R_s (v_d - E_d') - X' (v_q - E_q')] \tag{4.5}$$

$$i_q = \frac{1}{R_s^2 + X'^2} [-X' (v_d - E_d') + R_s (v_q - E_q')] \tag{4.6}$$

$$T_e = E_d' i_d + E_q' i_q \tag{4.7}$$

$$T_m = T_{mo} (A\omega_m^2 + B\omega_m + C), \; A\omega_{mo}^2 + B\omega_{mo} + C = 1 \tag{4.8}$$

$$\frac{d\omega_m}{dt} = \frac{1}{2H} (T_e - T_m) \tag{4.9}$$

$$T_o' = \frac{X_r + X_m}{\omega_o R_r} \text{ (transient open-circuit time constant)} \tag{4.10}$$

$$X = X_s + X_m \text{ (rotor open-circuit reactance)} \tag{4.11}$$

$$X' = X_s + \frac{X_m X_r}{X_m + X_r} \text{ (blocked rotor short-circuit reactance)} \tag{4.12}$$

In these equations, voltages and currents are in terms of d and q axis components referred to a synchronously rotating reference frame. In the first two equations the variable *s* is slip. The companion book [9] provides a rigorous derivation of the equations for the various types of induction motors.

More detailed induction motor models are available in transient stability programs such as EPRI's Extended Transient Mid-Term Stability Program (ETMSP) [10]. These may be necessary for large individual motors in an industrial plant, or for power plant auxiliary motors. For smaller motors, however, an even simpler model involving the steady state equivalent circuit and the inertial differential equation (Equation 4.9) is often adequate.

The LOADSYN program requires the user to specify the percent of each load class (residential, commercial, industrial, etc.) for each bus, area, or zone. The program then determines the load class composition and the load characteristics by component (Table 4-1). For load class composition, the user can enter data for various seasons and load levels (on-peak, off-peak) or can use one of several built-in load composition data bases. The user can specify that loads above a certain value include dynamic motor models—the program will determine the data sets for an aggregated motor model.

Figure 4-7 shows the load modeling structure.

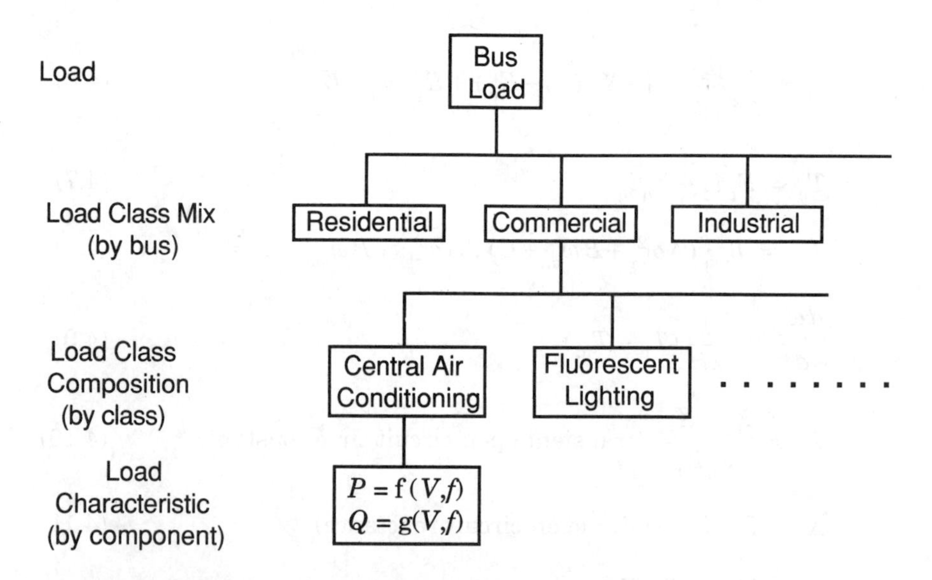

Fig. 4-7. LOADSYN modeling structure [3].

In large-scale studies, the user normally specifies the forecasted active and reactive loads at the high voltage side of a bulk power delivery substation. Using the power factor data of the load components, LOADSYN determines the total reactive load of the components. The difference between the user-specified reactive load and the LOADSYN-computed reactive load is due (at least partly) to distribution system reactive losses and reactive compensation. LOADSYN adds reactive compensation at the high voltage bus to match the user-specified reactive load.

LOADSYN aggregates the load voltage/frequency characteristics of the load components for classes and converts them to load models recognized by several power flow and stability programs. The model often used is a combination of constant impedance, constant current, and constant power. The conversion is a curve-fitting process. LOADSYN aggregates motor load providing dynamic models for stability programs [5]. Thresholds may be set for use of dynamic models based on size of load or distance from fault.

The interpretation of the Pv and Qv terms in Equations 4.1 and 4.2 and Table 4-1 is that Pv is the ΔP for a ΔV and Qv is the ΔQ for a ΔV. We can show this using a two-term binomial expansion of Equation 4.1 with frequency constant:

$$\frac{P_0 + \Delta P}{P_0} = \left[\frac{V_0 + \Delta V}{V_0}\right]^{Pv}, \quad 1 + \frac{\Delta P}{P_0} = \left[1 + \frac{\Delta V}{V_0}\right]^{Pv} \cong 1 + Pv\left[\frac{\Delta V}{V_0}\right]$$

$$Pv = \frac{\Delta P / P_0}{\Delta V / V_0}, \quad Pv = \frac{\Delta P}{\Delta V} \text{ with } P_0 = V_0 = 1 \text{ pu}$$

Distribution feeder representation. Figure 4-8 shows the typical model for large-scale power flow and stability simulations. Figure 4-9 shows a model with an equivalent representation of the distribution system. Comparing the two models, we note that the simple model has the following shortcomings [3]:

- The load (including compensation) responds to voltage and frequency changes at the high voltage bus rather than at the load itself.
- Reactive losses and compensation effects are direct functions of the high side bus voltage; they are really more dependent on current in feeders and transformers.
- For power flow and longer-term dynamics, tap changing effects are not represented.

Example 4-2. During voltage decay or during a voltage swing, voltage at a bulk power delivery high voltage bus drops from 1.05 per unit to 0.9 per unit. With no regulation, calculate the voltage at the utilization point assuming a reactance of 15% on the load base. (The reactance comprises

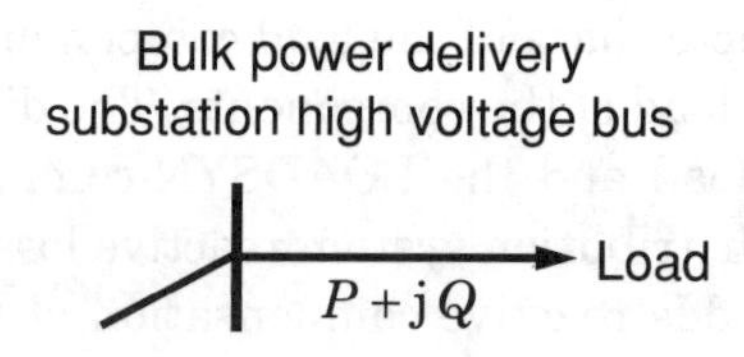

Fig. 4-8. Typical large-scale simulation model using LOADSYN.

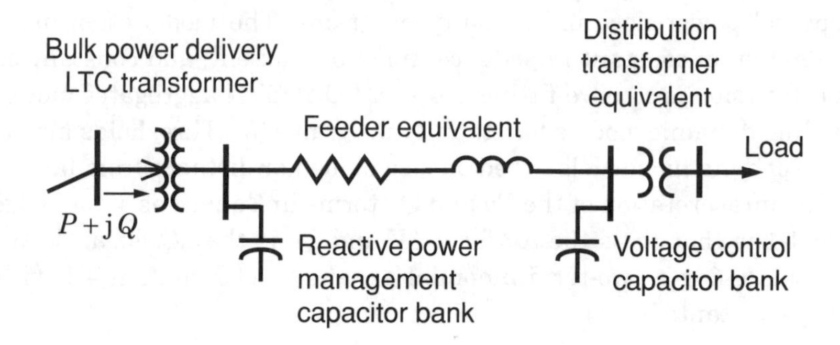

Fig. 4-9. Load model with distribution system equivalent.

10% bulk power delivery transformer reactance, 3% distribution transformer reactance, and 2% feeder reactance.) The voltage at the load, V, is initially 1.0 per unit. The 1.0 per unit load is a motor with $Pv = 0$ and $Qv = 2$. Capacitive compensation is used to achieve 1.0 per unit voltage. Calculate the voltage at the load when the source voltage, E, drops to 0.9 per unit.

Solution: Equations 1.1 and 1.2 are used to calculate initial reactive power flow to the load.

$$\delta = \arcsin \frac{1 \times 0.15}{1.05 \times 1} = 8.2°$$

$$Q_r = \frac{EV\cos\delta - V^2}{X} = 0.26 \text{ pu}$$

Q_r is matched by the reactive load of the motor (0.75 per unit for 0.8 power factor) and the capacitive compensation. The net reactive load is a reactance since $Qv = 2$. With the source voltage drop from 1.05 to 0.9 per unit, the following equation matches the load reactive power to the delivered reactive power.

$$0.26V^2 = \frac{0.9V\cos(\arcsin(0.15/0.9V)) - V^2}{0.15}$$

The voltage at which the left hand side equals the right hand side can be solved by trial and error using a spreadsheet program. The solution is $V = 0.8492$ per unit. The voltage gradient is not much different from the initial 5% gradient.

Our next example is more dramatic.

Example 4-3. Reference 3, Volume 1, computed power flows for the simplified distribution system shown on Figure 4-10. The computations are for a range of source voltages and for three types of active loads (constant power, constant current, and constant impedance). The reactive load is constant impedance. Figure 4-11 shows the results, which you should examine.

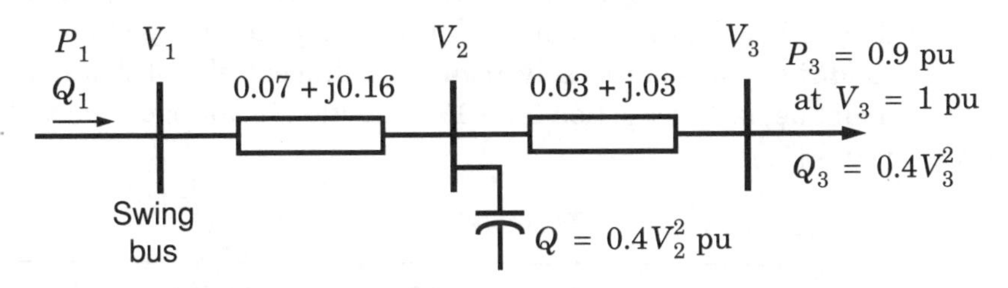

Fig. 4-10. Example 4-3: Distribution system representation.

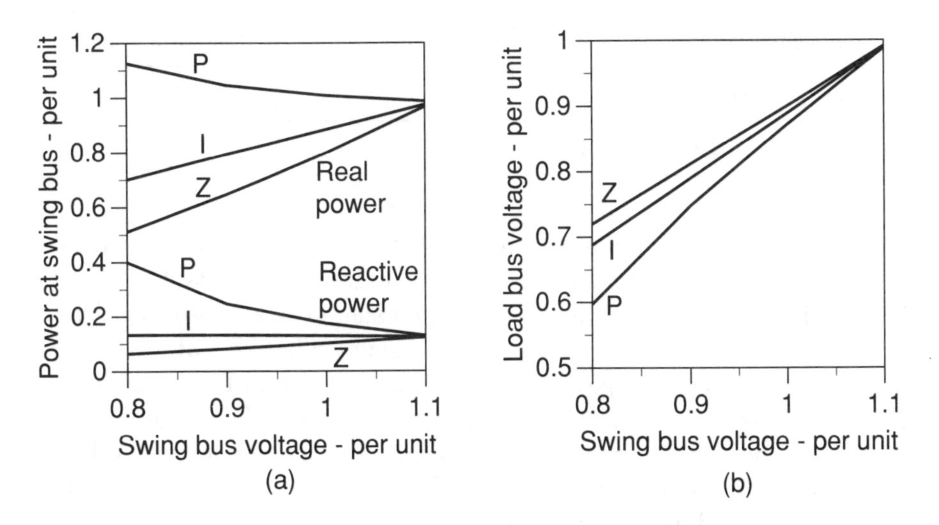

Fig. 4-11. Example 4-3: Plots of variables versus source voltage for three types of active load.

Voltage reduction experience. Many utilities use intentional voltage reduction to obtain load reduction during times of generation or transmission capacity shortages. Since the oil embargo of the 1970s, utilities have evaluated and implemented voltage reduction to conserve energy (conservation voltage reduction)—often after pressure from regulatory agencies. Many utilities have also used voltage reduction as an emergency measure to improve voltage stability. For example, the 1988 North American Electric Reliability Council (NERC) System Disturbance Report listed eight days of voltage reduction in the eastern U.S. during the summer of 1988.

Voltage reduction is accomplished through tap changing transformers and distribution voltage regulators.

Many field measurements of the effectiveness of voltage reduction have been reported in the literature; references 11–13 are a sampling. To be effective, reduction of voltage-sensitive load must overwhelm the loss of load diversity (i.e., heaters staying on longer). Experience has generally been favorable, although savings may not always be great enough to justify the capital additions to more closely regulate voltage. Table 4-2 summarizes the percent demand reduction for a 1% reduction in voltage from tests by three utilities [14]:

Table 4-2

Company	Residential	Commercial
American Electric Power 1	0.80%	0.78%
American Electric Power 2	0.90	0.86
Consumers Power Company	0.83	1.38
San Diego Gas & Electric Co.	1.14	0.08

Voltage reduction is a tool to improve voltage stability. Load characteristics and composition data are needed to evaluate the benefits during critical heavy load periods. The voltage reduction should be as close as possible to the loads and downstream from shunt capacitor banks. Often the percent reduction in reactive power load is much more than the reduction in active load. This is because distribution transformers and some motors are operated in magnetic saturation at normal voltage. Following voltage reduction, constant energy active load will be restored after some minutes by thermostatic and manual control.

Distribution automation facilitates implementation of voltage reduction. Reference 15 describes application at Virginia Power Company.

The active and reactive power voltage sensitivities we have described are for small voltage changes. For voltage stability, the performance of certain loads at quite low voltage is also important.

Discharge lighting. About one-third of commercial load is lighting [16] — largely fluorescent. Fluorescent and other discharge lighting has a voltage sensitivity Pv in the range 1–1.3 and Qv in the range 3–4.5. At voltages between 65–80% of nominal they will extinguish, but restart (with a hysteresis characteristic and one or two seconds time delay) when voltage recovers. For a large population of lamps, a single-value function (no hysteresis) model is used: between 75% and 65% voltage the discharge lighting load is ramped to zero [17,18].

There are new fluorescent lighting controls that rapidly vary lamp output to complement natural lighting. These controls correct for voltage variations, resulting in a constant power load characteristic.

Induction motors. About 57% of the U.S. electricity consumption goes to power motors, mostly integral horsepower three-phase induction motors [17]. About 78% of industrial sector energy use is for motors; the corresponding values for residential and commercial sectors are 37% and 43%. Induction motors consume about 90% of total motor energy.

Most of residential and commercial motor use is for the compressor loads of air conditioning and refrigeration. Compressor loads require nearly constant torque at all speeds, and are the most demanding from a stability viewpoint. Pumps, blowers and fans, and compressors account for more than half of industrial motor use [19].

Surveys cited in reference 19 found that motors are commonly oversized, with perhaps half of all integral horsepower motors operating at less than 60% of rated load. Large industrial motors and motors supplied as part of packaged equipment, however, are usually properly sized.

The energy consumption patterns are not representative of peak summertime load conditions when heavily loaded air conditioners comprise a much larger share of motor load. Domijan et al. [20] state that 30% of energy consumption in Florida is for air conditioning.

The steady-state active power drawn by motors is fairly independent of voltage until the point of stalling. Motor reactive power is more sensitive to voltage levels and other effects. As voltage drops, the reactive power will first decrease, but then increase as the voltage drops further. On Table 4-1 the Qv cited for motor loads varies between 0.4 and 3.5. Values of Qv greater than 2 probably reflect operation with motor core saturation.

Many motors, and appliances containing motors, were tested as part of EPRI-funded research on conservation voltage reduction. Figure 4-12 shows typical results for a 20 horsepower motor [11]. As expected, the

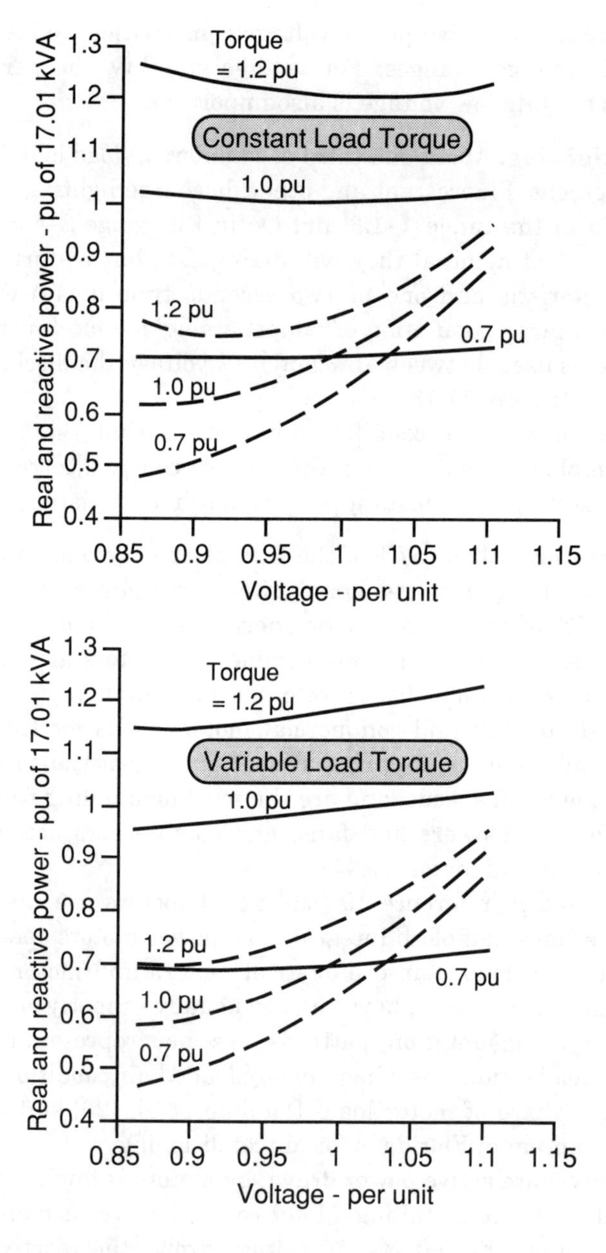

Fig. 4-12. Static active (solid lines) and reactive power (dashed lines) curves versus voltage for a 20 horsepower, 230-volt, three-phase induction motor. Top curves for constant torque; bottom curves for torque proportional to the square of the speed [6, Volume 1].

curves show that constant torque loads and heavy loadings are more onerous for voltage stability; the motor is stall-prone. The reactive power voltage sensitivity decreases with increased loading. From the slopes of the curves, values of Qv can be determined. For constant torque loads, the values (around rated voltage) are about 1.8, 1.4, and 1.0 for loadings of 0.7 per unit, 1.0 per unit, and 1.2 per unit, respectively. For variable torque loads, the corresponding slopes are 1.9, 1.5, and 1.3.

Air conditioning and heat pumps. Voltage instability is likeliest during very high load levels caused by extreme temperatures. For summertime conditions, a major part of the load will be air conditioners. Figure 4-13 plots test data from EPRI-funded research [11]. The figure shows typical air conditioner static performance as a function of temperature and volt-

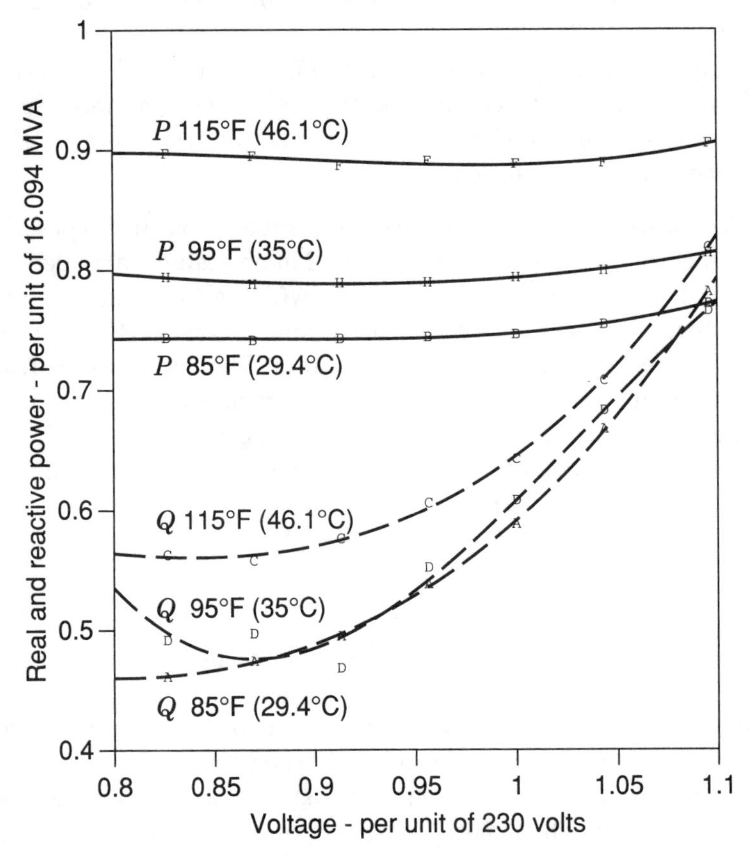

Fig. 4-13. Static active and reactive power curves versus voltage for 119,000 BTU, 208/230 volt, three-phase central air conditioner [11, Volume 2: Appendixes]. Data is curve fitted; some data for reactive power at 95° is suspect.

age. Heat pumps have similar characteristics, except for the cooling mode at very cold temperatures when the supplementary resistance heating may dominate. Related to Figures 4-12 and 4-13, reference 11 provides test results for other loads.

We note from Table 4-1 that air conditioners have low inertia constants (H) and thus are prone to stall. Reference 21 reports recent tests of recovery of air conditioner motors following faults. The tests showed that air conditioners will decelerate and stall at fault voltages below about 60%—assuming five-cycle or longer fault clearing time. With slower fault clearing, air conditioners may stall at higher fault voltages. After source voltage returns to normal, stalled motors will not recover until compressor pressure bleeds off. Thermal overload protection will disconnect the air conditioner in 3 to 30 seconds. Large commercial air conditioners have undervoltage relays, however, that trip the units within five cycles after voltage drops below 70% [21].

The July 23, 1987 Tokyo blackout was partly due to characteristics of new, electronically-controlled air conditioners (load commutated inverter). [22]. The characteristics of the new equipment is even more unfavorable than conventional air conditioners.

Induction motor torque-speed curves. Torque-speed curves describe the stall characteristics of induction motors. Motor stalling also depends on inertial and flux dynamics which, along with network and disturbance characteristics, determine motor speed and voltages. Motor electrical torque is proportional to the voltage squared. This can be seen from the equivalent circuit shown of Figure 4-5. Neglecting the shunt magnetizing path, the motor current is the terminal voltage divided by the circuit impedance. The electrical torque is the air gap power (less the rotor copper loss) divided by the motor speed:

$$T = \frac{1}{\omega_s} i^2 \frac{R_r}{s} \tag{4.3}$$

Since torque is proportional to current squared, it is also proportional to voltage squared.

Figure 4-14 shows torque-speed curves, as a function of voltage and temperature, for a residential central air conditioner compressor motor. The constant torque mechanical characteristic shown is the most demanding. Stalling occurs when mechanical torque exceeds available electrical torque, resulting in deceleration. For compressor loads (air conditioners and refrigerators), restarting or reaccelerating following a brief outage or voltage dip may not be successful until compressor pressure bleeds off [21–23].

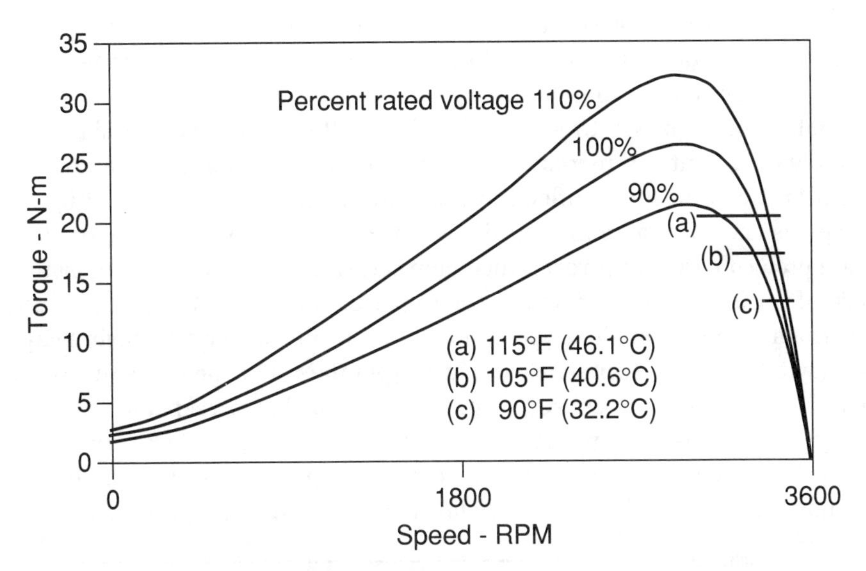

Fig. 4-14. Torque-speed curves for 5 HP single-phase residential central air conditioner compressor motor. Source: General Electric Company.

Considering torque-speed curves such as Figure 4-14, the improved stability of mechanical loads with torque varying as the square of the speed can be visualized. A speed-squared torque relation characterizes fan and pump loads.

Induction motor dynamics. The dynamic characteristics of motors are critical for voltage decay or voltage dips below about 0.9 pu. At sustained low voltages between about 0.7 and 0.9 pu, many motors will stall and draw large amounts of reactive power. Stalling of one motor may cause nearby motors to stall. A related problem is reacceleration of motors following faults.

Besides the individual motor data of Table 4-1, The LOADSYN research developed aggregated motor data sets using frequency domain techniques [3,5].

Think about the equivalent circuit of induction motors (Figure 4-5). Following a disturbance, a motor will first act as an impedance load— the slip, s, cannot change instantly, only after inertial dynamics lasting a few tenths of a second. This "impedance jump" response to a step change in voltage [24] will be pointed out in simulation examples in Chapter 6.

Example 4-4. This example is based on Sekine and Ohtsuki [24]. Consider the *P–V* curve of an induction motor fed from a weak power system. Assume the motor: (1) is at a normal operating point, and (2) is transiently on the bottom side of the *P–V* curve. Analyze the effect of energizing a capacitor to stabilize the motor. Will the voltage increase or decrease? What will be the final operating point?

Solution: Figure 4-15 shows that low voltage (i.e., bottom side of the *P–V* curve) results in increased slip. Figure 4-16 shows the system under consideration and a simplified induction motor equivalent circuit. Figure 4-17 shows the corresponding static (constant motor power) *P–V* curves and the motor constant slip resistance characteristics. Point A is the assumed normal initial operating point on the upper part of the *P–V* curve, and point B is the assumed transient operating point on the bottom side before capacitor switching (the transient operating point is at a power equal to the mechanical power). Because of the initial impedance response, capacitor energization causes an immediate jump from point A to A′ or from point B to B′. In either case, the electrical torque (or power) is greater than the mechanical torque. The motor accelerates, reducing the slip to point C.

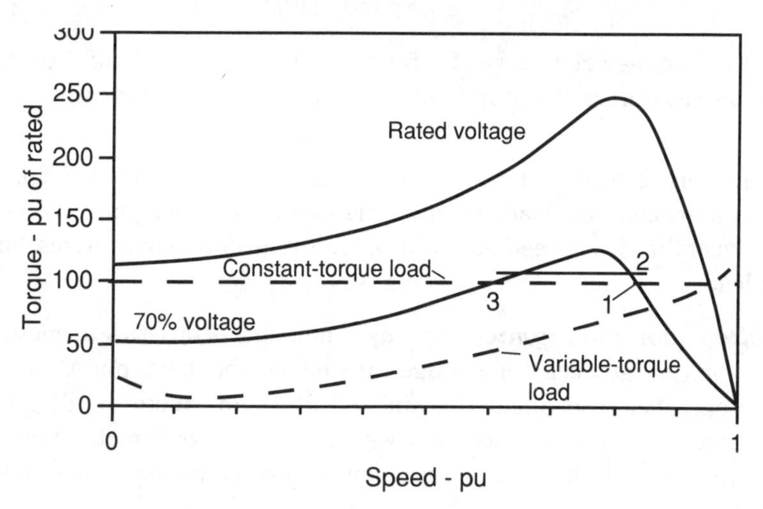

Fig. 4-15. Torque-speed curve showing reduction in slip caused by a reduction in voltage.

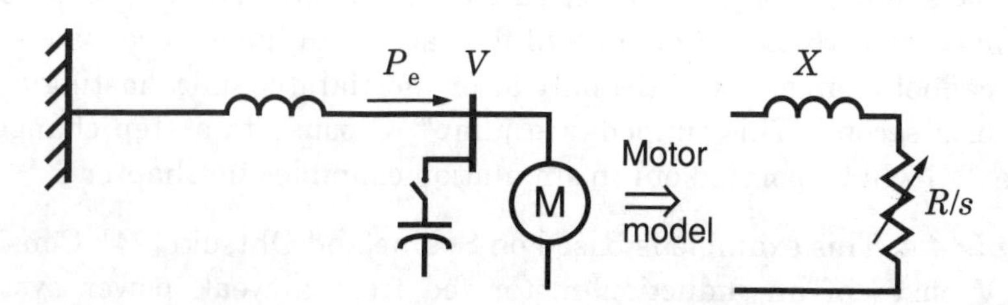

Fig. 4-16. Example study system and simplified induction motor model.

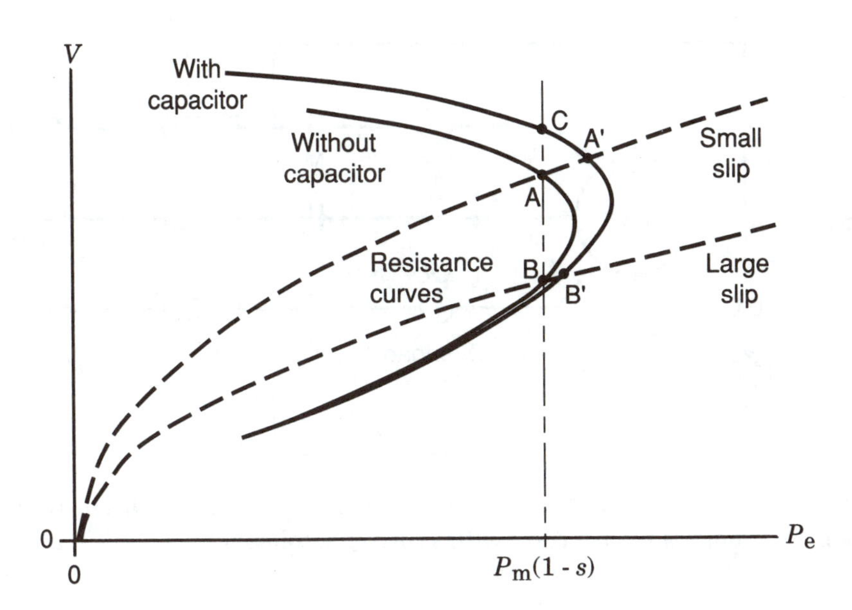

Fig. 4-17. *P–V* curve showing how capacitor energization will result in operation at point C for initial operation at either point A or B [24].

If voltage decays to a value less than point B and if point B′ (after capacitor switching) is less than the initial power, the motor decelerates rather than accelerates. Motor stalling and voltage collapse follows.

Referring back to Figure 4-15, we can further describe conditions for motor stability. Consider operation at point 1 with low voltage and constant torque load. A small increase in mechanical load as shown by the solid line causes the motor to decelerate to new stable operating point 2. At a slip smaller than the slip at maximum torque (point 3), however, the deceleration caused by a small increase in mechanical torque will result in motor stalling. Momentary operation at point 3 could be due to deceleration during a fault (short circuit) followed by insufficient voltage recovery after line tripping to clear the fault.

Again referring to Figure 4-15, motor instability and stalling for a gradual reduction in voltage occurs when the motor maximum torque is reduced to where there is not an intersection with the mechanical load characteristic. In either case, motor instability and stalling could lead to voltage collapse.

Motor controls and low voltage contactor releases. Equally important as motor dynamics and possible stalling are the protection and undervoltage releases of motors. Many, if not most, industrial motors have starter

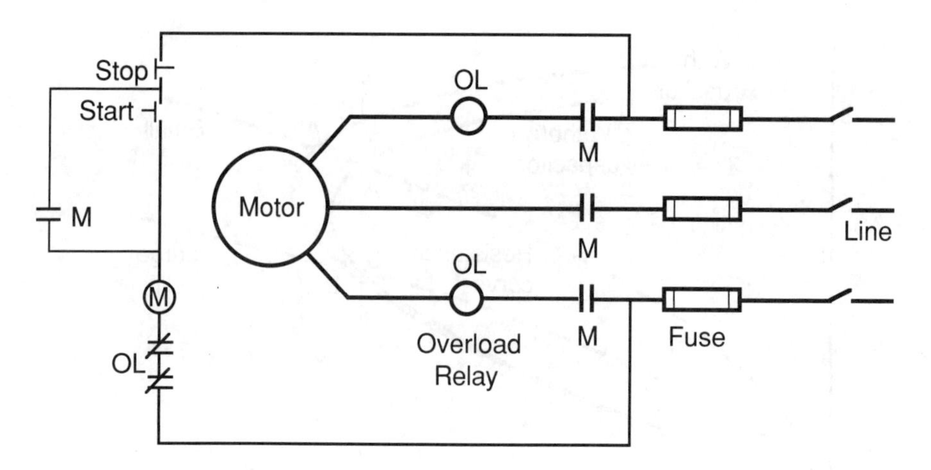

Fig. 4-18. Motor starter circuit showing contactor release. Often, control power is through a transformer with 110-volt secondary.

controls with ac contactors similar to Figure 4-18. Motors will immediately trip off line when the phase-to-phase voltage energizing the "M" relay drops low enough. The dropout characteristics are variable. Dropout voltage ranges from about 30% voltage to over 65% voltage and the dropout time ranges from less than a cycle to as long as ten cycles [25]. In the field, contactor dropout has occasionally been experienced at even higher voltage (80–90% voltage). This can be caused by poor maintenance, contamination, or incorrect application (for instance a 600-volt starter on a 440-volt motor).

There are several alternatives to the circuit shown in Figure 4-18 that reduce unnecessary tripping of critical motors. These include mechanical latch "M" relays and control power from a source other than the power system. Starters are available that will reconnect the motor following a momentary voltage dip. Very large motors have circuit breakers and relays which will trip the motor only if damage is imminent.

Small motors on appliances and tools (refrigerators, single-phase air conditioners, etc.) normally have only thermal overload protection. Voltage recovery problems owing to single-phase air conditioners stalling on slow-clearing subtransmission and distribution faults have been reported [21]. Before the stalled air conditioners are removed by thermal protection (several seconds), voltage in a load area can be dragged down. Stalled motors will draw current four to six times normal, prolonging the voltage dip. Ground relays may trip lines because of the unbalanced loading.

Because of the many unknowns regarding motor control and protection, disconnection of load should not be relied on to prevent voltage instability.

Adjustable speed drives. Motors using power electronics for variable speed control are becoming common. If controlled rectifiers are used, power factor will be improved for low voltage as the firing angle is reduced to maintain dc voltage [26]. Experience has been that adjustable speed drives will drop off line at about 90% voltage. This may eventually change as manufacturers respond to complaints about nuisance trips.

Synchronous motors. Synchronous motors are normally used in high power (megawatt-level) applications. They are more complex and expensive than induction motors, but are also more efficient. Use of excitation control to regulate voltage results in favorable characteristics for voltage stability.

Modeling is similar to synchronous generators except that a mechanical load model such as Equation 4.8 is required.

Electronic power supplies. Regulated power supplies on computers, and other electronics will provide constant dc voltage down to around 90% voltage [27]. Below this voltage the power will fall and the equipment may not operate properly. Figure 4-19 shows an envelope of voltage tolerance that is representative of the present design goal of a cross section of the electronic equipment manufacturing industry [28,29].

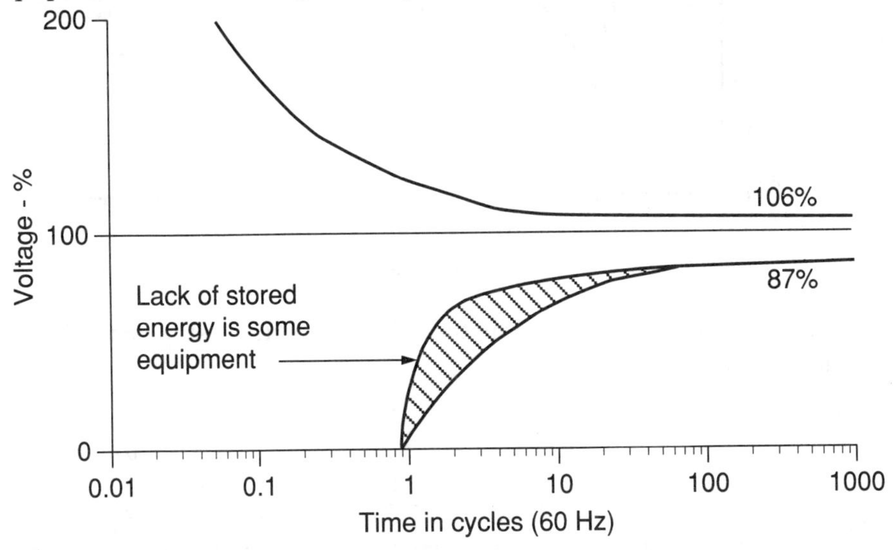

Fig. 4-19. Typical voltage tolerance envelope for electronic power supplies, ANSI/IEEE Std 446-1987. ©1987 IEEE.

Constant energy loads and load diversity. Loads such as space heating, water heating, industrial process heating, and air conditioning are controlled by thermostats, causing the loads to be constant energy. For heating loads, low voltage results in loss of load diversity since individual loads stay on longer. Over time, the aggregated load changes from resistive to constant power. Graf [30] studied the response of large numbers of constant energy loads to voltage drops and developed a single time constant model. Figure 4-20 shows the response of many loads to a step reduction in voltage. The response time constant is around four minutes. For very large voltage drops, the load is not restored, indicating all loads are on continuously and the required energy cannot be satisfied. For the 10% voltage reduction there is a small amount of overshoot.

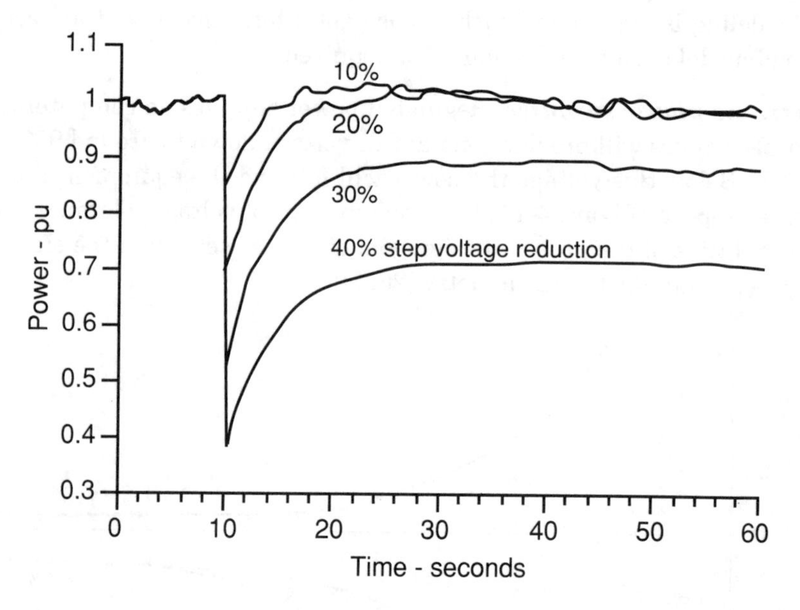

Fig. 4-20. Response of 10,000 constant energy loads to step reduction in voltage [30].

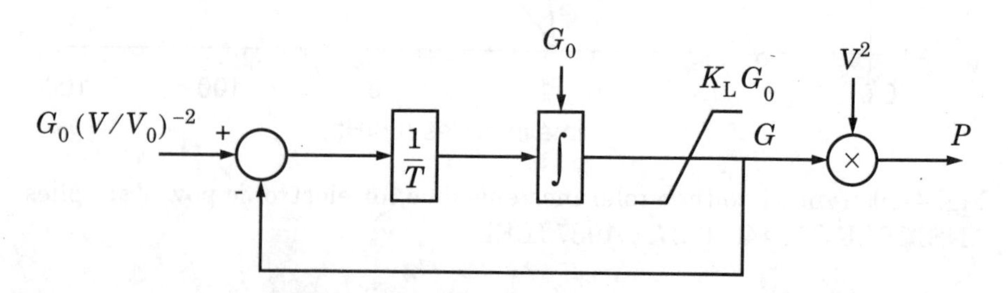

Fig. 4-21. Single time constant model of response of many constant energy loads to voltage changes [30]. G is load admittance, G_0 is initially connected load conductance. $K_L\,G_0$ is conductance with all loads on.

Figure 4-21 shows a single time constant model.

Field measurements at two substations in Sweden showed similar active load recovery with time constants of only two–four minutes [31,32].

Electric space heating may be a large load during cold weather. The effective time constant for loss of diversity becomes shorter during cold weather depending on factors such as temperature, wind, and building thermal time constants. The heater-on cycle time will be longer and the off cycle time will be shorter.

Example 4-5. Consider four thermostatically-controlled electric space heating loads of 1 pu power each. The weather conditions are such that, for 1 pu voltage, they are on for four minutes and off for four minutes. The on/off cycles of the four heaters are initially symmetrically distributed so that two heaters are on at any point in time. The voltage suddenly drops to 0.894 pu so the heater power drops to 0.8 pu. Calculate the new on and off times for constant energy and sketch the time response.

Solution: The average power per heater is 0.5 pu. This must be maintained for constant energy. Following the voltage reduction, the off time will remain at four minutes. The on time will increase. We can calculate:

$$\text{avg } P = 0.5 = \frac{0.8\,(t_{on}) + 0\,(t_{off})}{t_{on} + t_{off}}, \; t_{on} = 6.67 \text{ minutes}$$

Figure 4-22 shows the response. The load initially drops by 20% but in two minutes temporarily overshoots to 120% of initial load. With more loads, the response will approach that shown on Figure 4-19.

Air conditioning and other compressor loads are constant energy load. However, since they are nearly a constant power load before thermostatic control, there will be little change in load diversity with voltage changes.

Generation in load area. Relatively small non-utility generation (cogenerators associated with industrial loads or independent power producers) may be embedded in a load area. The larger synchronous generators (and synchronous motors) with automatic voltage regulators will improve voltage stability. Excitation control on smaller synchronous generators connected to distribution networks may instead regulate power factor so as to not interfere with utility voltage regulation [33].

Many non-utility generators are induction generators with shunt capacitor compensation. Performance is similar to induction motors.

Induction generators are simpler and cheaper. Synchronous generators, however, are more efficient and are usually used for the larger power ratings.

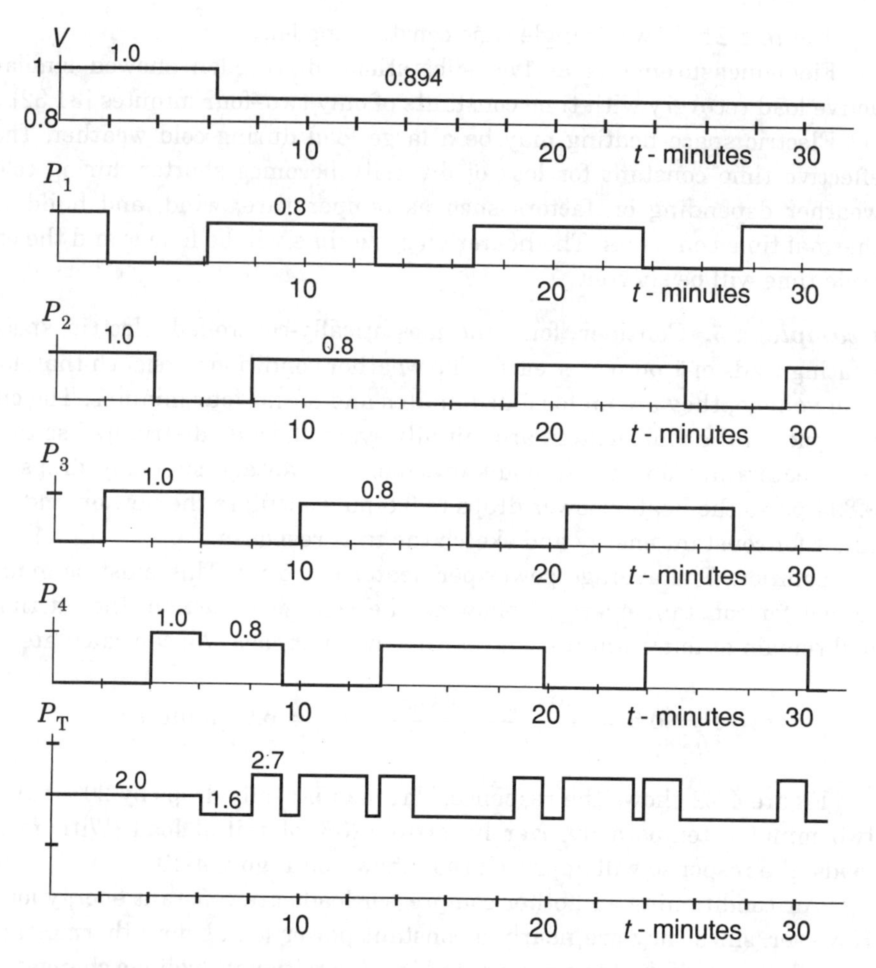

Fig. 4-22. Example 4-5: response of four thermostatically controlled heaters to voltage reduction. Heater off time is 4 minutes. Heater on time is 4 minutes initially, and 6.67 minutes after voltage reduction.

Representing embedded generation in simulations is a challenge. Network reduction software can be used to develop equivalents for subtransmission networks which include non-utility generators of significant size.

4.3 Reactive Compensation of Loads

The secondary or distribution side of bulk power delivery substations often includes shunt capacitor compensation. The capacitor banks are generally for "reactive power management" rather than for direct voltage control. For example, if the load drawn from the main transformer is high, capacitors are switched on. Heavy load means relatively high reactive demand and reactive losses. The capacitor banks are relatively large and at least some banks are switchable. The capacitors also release transformer capacity.

One approach to reactive power management is to minimize reactive transfer between voltage levels. This supports the general principle of supplying compensation close to reactive power consumption.

Control variables include current, reactive power, power factor, time, temperature, and combinations. Current and combinations (which usually includes voltage) are the most used. Capacitor control must be coordinated with tap changer control. Control variables other than voltage, such as current, facilitates coordination.

Distribution feeders often have capacitors distributed along their length. Switched capacitors near the end of feeders are usually voltage controlled.

With distribution automation, the future will probably see more centralized control of substation capacitor banks and feeder capacitors to optimize operation of an entire area.

Industrial load usually includes shunt capacitor compensation. This is because of lagging power factor induction motors. Depending on the load, harmonic filters (capacitive at fundamental frequency) may be needed.

Static var compensators. Transmission applications of static var compensators are described in Chapter 3. Static var compensators, however, were first developed for large fluctuating industrial loads such as arc furnaces and rolling mills. Sizes range from about 25–100 MVAr. The devices are effective and many installations are in service. Current or reactive power rather than voltage is often controlled. Miller [34] devotes a chapter to reactive compensation of arc furnaces.

SVCs (as small as one MVAr) have been applied at smaller loads. Reduction of voltage dip during starting of large motors is one application [35]. Application at fluctuating loads include mining loads, saw mills, paper mills, and induction furnaces. Single phase loads include automatic welders and electrified railroads.

Figure 4-23 shows an eleven step, 2475-kVar SVC connected to a 25-kV distribution network [35]. The SVC, consisting of five 450-kVAr TSCs and one 225-kVAr TSC, operates at 600 volts. All 600 volt equipment is indoors.

Gate-turn-off thyristor (GTO) based static var compensators are being developed by SVC manufacturers. These will likely first be applied on distribution systems. For voltage stability, benefits are that outside control limits current output is constant rather than proportional to voltage. Performance is similar to synchronous condensers [36].

Fig. 4-23. Distribution network SVC rated 2475-kVAr. *ABB*.

Series capacitor compensation. Series capacitors are sometimes used in subtransmission and distribution circuits. Reference 37 and 38 describe applications. Often, series capacitors can be thought of as a voltage booster—an instantaneous, continuous voltage regulator. Figure 4-24 shows the voltage boosting obtained with a lagging power factor load. Increased current or reduced power factor increases the boost—exactly when needed. The instantaneous response is important when load fluctuations cause voltage flicker. Solving voltage flicker is a primary application of distribution series capacitors.

There are difficulties with series capacitors including ferroresonance in transformers and subsynchronous resonance during motor starting [39,37]. Some modern distribution series capacitor systems use metal oxide varistors and other circuits to eliminate spark gaps and SSR damping resistors [40]. Figure 4-25 shows a recent distribution series capacitor installation.

4.4 LTC Transformers and Distribution Voltage Regulators

In Chapter 2, we describe tap changing as a principal mechanism leading to voltage instability. Following a disturbance, the voltage sags provide temporary relief; in less than a minute, however, tap changing equipment

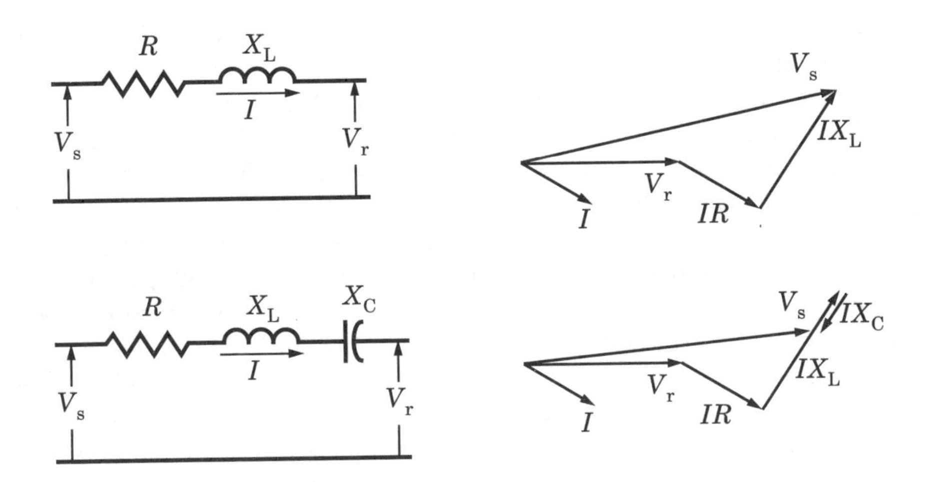

Fig. 4-24. Phasor diagram showing series capacitor voltage boost with lagging power factor load [37].

Fig. 4-25. Distribution network series capacitor. *General Electric Company.*

starts to restore the load-side voltage, and thus the load. We will now examine tap changing.

Bulk power delivery load tap changing (LTC) transformers and distribution voltage regulators act similarly in regulating load-side voltage. A voltage relay monitors load-side voltage. If voltage drifts, or jumps, outside a deadband (bandwidth is typically 2 volts on 120-volt base or ±0.83%), a timer relay is energized. (An inverse-time voltage regulating relay may be used in older installations rather than an instantaneous voltage relay and timer.) If the relay times out (after tens of seconds), the tap changing mechanism will be energized and tapping takes place until the voltage goes in-band, or until the maximum or minimum tap is reached. Once in-band, the voltage relay and timer mechanism is reset. In the course of a voltage collapse, tap changers on individual transformers or voltage regulators may reset several times, slowing the voltage collapse process.

Bulk power delivery LTC transformers and distribution voltage regulators usually have ±10% tap range consisting of thirty-two steps of 5/8% each.

Time delay before tapping is adjustable—a range of 10–120 seconds is common. Thirty or sixty seconds delay is typical. For a very large voltage drop, the time to tap from neutral to full boost (timer delay followed by sixteen steps) is often around two minutes.

Practices of utilities outside North American are often different. In Europe, constant or inverse timing (5–120 seconds) is used between each tap step. With a long time delay between each step, voltage decay will be much slower, allowing more time for corrective measures.

Figure 4-26 shows a 12/16/20 MVA bulk power delivery LTC transformer. Figure 4-27 shows a single-phase voltage regulator.

Calovic [41] provides a detailed model for tap changers. The model, shown on Figure 4-28, is adapted for use in the EPRI Extended Transient Mid-Term Stability Program (ETMSP).

Table 4-3 shows test results by a utility to determine time for tapping from neutral to full boost. Note that thirty second delay, with about two minutes for full raise is about average. (For a large voltage drop during heavy load conditions, fewer than sixteen steps would usually be available.)

Line drop compensation. Particularly for individual feeder regulators, line drop compensation is often used to regulate a point farther down the feeder. Figure 4-29 shows a circuit with line drop compensation to regulate a point $R + jX$ primary ohms down the feeder [37]. Note that the capacitor current is subtracted in the current transformer circuit. The capacitor would be switched by a different measurement (current, time, temperature,

Fig. 4-26. 12/16/20 MVA LTC transformer. *Cooper Power Systems.*

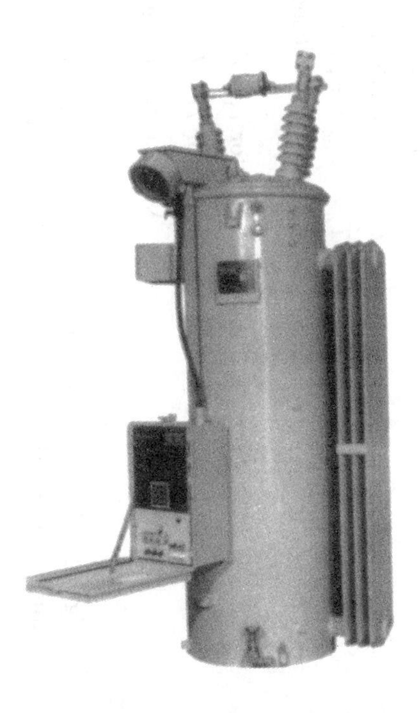

Fig. 4-27. Single-phase distribution voltage regulator. *Cooper Power Systems.*

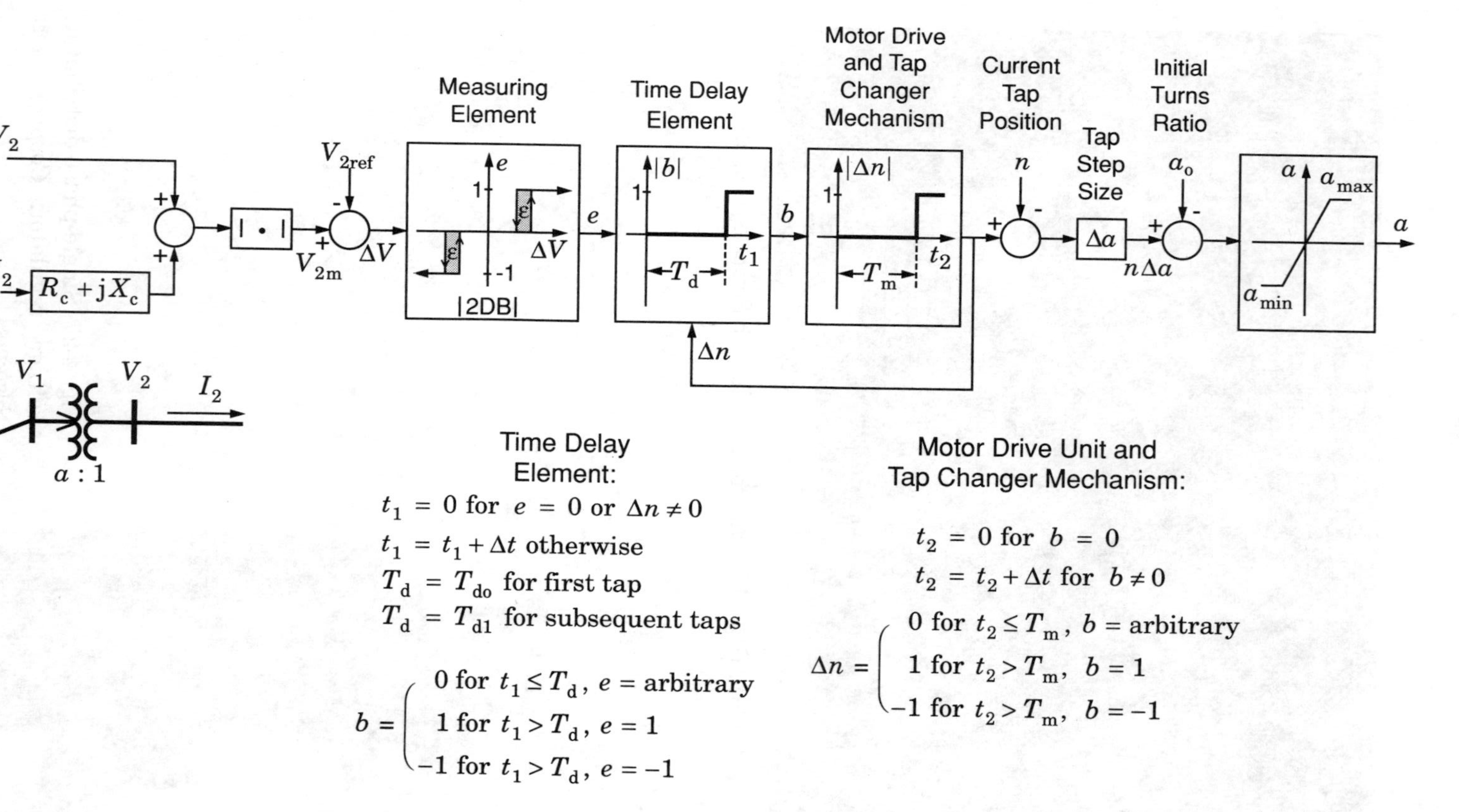

Fig. 4-28. Model for load tap changer transformer regulating secondary voltage (adapted from reference 41). For North American practice, set $T_{d1} = 0$. For intentional delay between each tap, set $T_{d1} = T_{do}$.

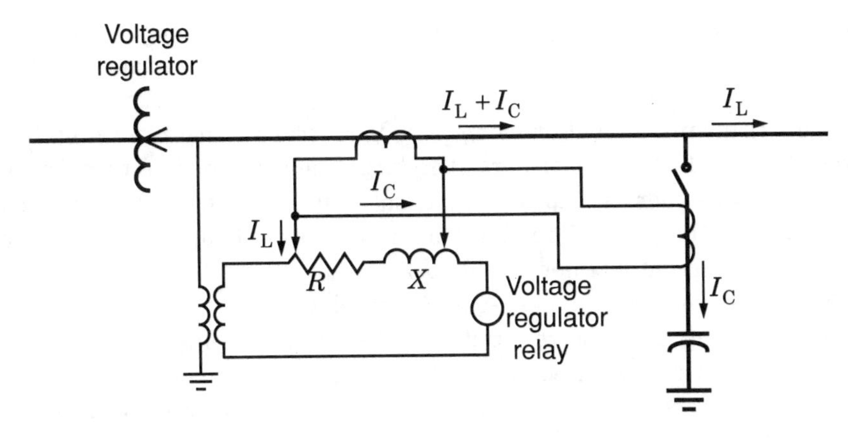

Fig. 4-29. Distribution voltage regulator with line drop compensation.

or voltage on generation side of regulator). Gönen [1] describes methods to set line drop compensation when there is distributed load between the regulator and the desired point of regulation.

Table 4-3

Substation	LTC type	Timer setting - sec	Steps to full raise	Time to full raise - sec
A, 115/12-kV	Manuf. A, 1	60	16	92
B, 115/12-kV	Manuf. B	30	16	158
C, 115/21-kV #1	Manuf. C	45	16	125
C, 115/21-kV #2	Manuf. C	45	16	125
C, 115/21-kV #3	Manuf. D, 1	45	16	109
D, 115/12-KV	Manuf. D, 2	30	16	104
E, 230/21-KV #1	Manuf. E	30	16	76
E, 230/21-KV #2	Manuf. E	30	16	76
E, 230/21-KV #3	Manuf. E	30	16	70
F, 115/12-KV #1	Manuf. D, 3	30	16	94
F, 115/12-KV #2	Manuf. D, 3	30	16	158
F, 115/12-KV #3	Manuf. A, 2	30	16	108

LTCs in series. Occasionally, two (or even three) automatically-controlled tap changers are in series. A bulk power delivery LTC transformer may serve long feeders with voltage regulators along the feeders. Another possibility is transmission/subtransmission LTC transformers serving bulk power delivery LTC transformers. The larger generation-side transformer should have the shortest tap changer time delay. For a small change in network voltage, the generation-side transformer would tap one step before the load-side regulator timer would operate.

For larger voltage changes, such as associated with voltage instability, coordination might not be realized. The distribution voltage and the load may overshoot its original value [42,43]. Because of the combined effect of both tap changers, distribution voltage will restore before the upstream voltage. The upstream LTC will continue to operate, causing distribution voltage overshoot.

Figure 4-30 shows a possible time response to a step reduction in source voltage. We assume the subtransmission LTC has a thirty second delay in tapping and that the bulk power delivery LTC has a forty-five second delay. We also assume each tap step takes five seconds. Once the voltage passes through the controller deadband, the timer is reset and must time out again before correcting for an overshoot. (We ignore the feedback of changes of load and of subtransmission/distribution losses on voltages.) Figure 4-30 shows a three step overshoot, corresponding to about 2% overvoltage (5/8% steps).

Besides increasing load above pre-disturbance levels, the voltage overshoot could cause voltage-controlled capacitors to trip off [42].

The load overshoot problem can be avoided by short time delay (10–30 seconds) on the upstream tap changer and a long delay (60–120 seconds) on the downstream tap changer. Referring to Figure 4-30, you may sketch responses on graph paper for other conditions such as changing the second LTC transformer time delay from forty-five seconds to say, thirty and sixty seconds.

LTC transformer equivalent circuit. The primary effect of tap changing on voltage stability is restoration of voltage-sensitive load that is reduced during voltage sag. For an ideal transformer, and an impedance load, the load is reflected to the high side by the square of the turns ratio as described in Chapter 2. For a real transformer with leakage impedance, there are additional effects [44].

Figure 4-31 shows a LTC transformer equivalent circuit. For tapping to raise secondary voltage, the primary-side shunt element is a "mathematical" reactor and the secondary-side shunt element is a "mathematical" capacitor. The secondary support depends on additional reactive power

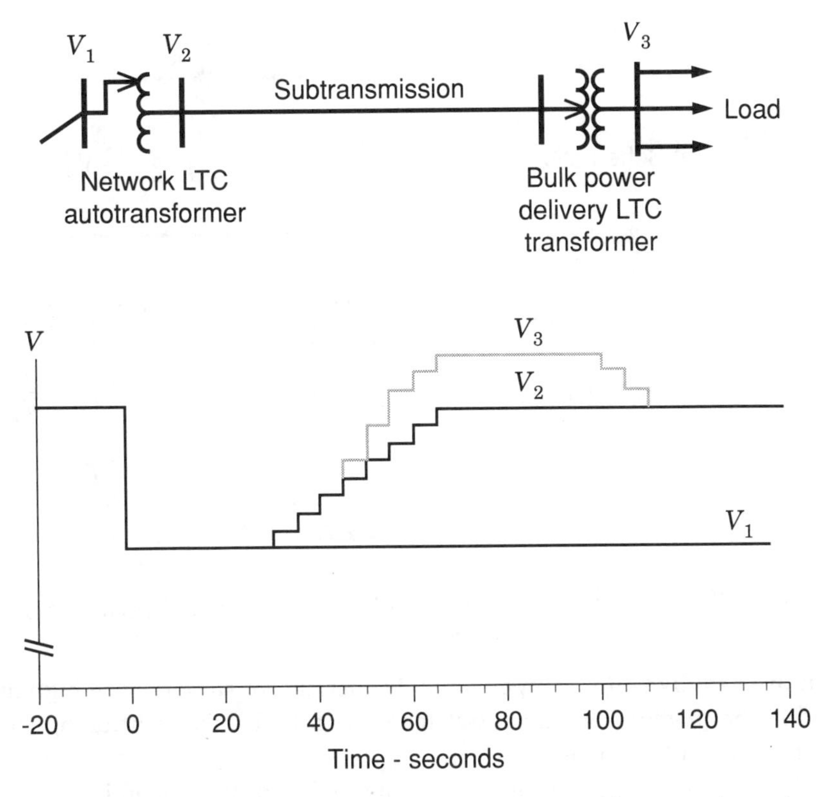

Fig. 4-30. Distribution voltage overshoot with two tap changers in series.

from the primary system. The primary or source system must be stiff enough to support this effect of tap changing.

Example 4-6. A 230/34.5-kV transformer has 10% leakage reactance (Y = -j10). Initially n = 1. Calculate the effect of tapping to raise the secondary voltage by 10%.

Solution: the off-nominal turns ratio, n, is 1.1. The series term (architrave) of the equivalent circuit becomes nY = -j11. The primary-side shunt term becomes $n(n - 1)Y$ = -j1.1. The secondary-side shunt term becomes $(1 - n)Y$ = j1.0

The shunt elements equate to a reactor of size $1.1V_1^2$ on the primary side and a capacitor of size $1.0V_2^2$ on the secondary side.

Effect of tap changing on shunt compensated loads. We have discussed how tap changing aggravates voltage stability by restoring load following voltage drops. This applies particularly to high power factor loads. Sometimes, however, voltage regulation by tap changing can improve voltage stability. This is true for lagging power factor, voltage-insensitive loads

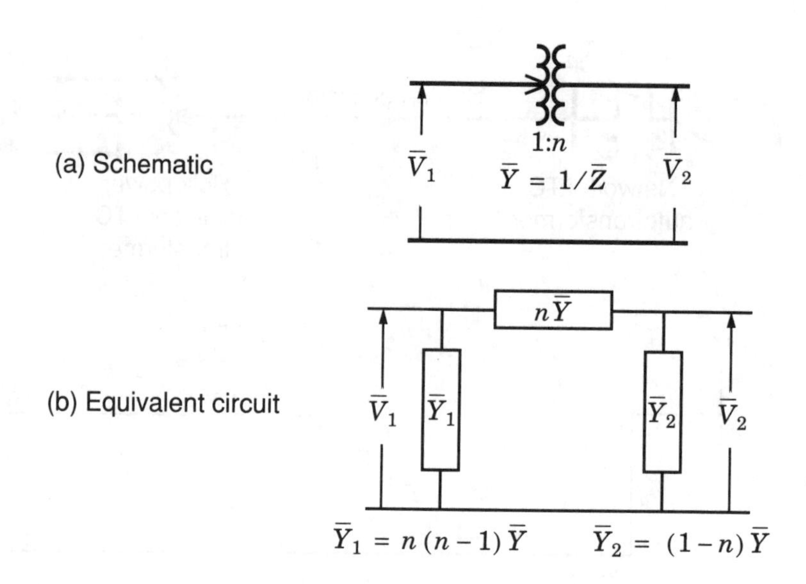

Fig. 4-31. Transformer equivalent circuit, n is off-nominal turns ratio.

that are heavily shunt compensated. Industrial consumers with high motor load is the primary example, but loads with high air conditioning component have similar properties.

The explanation is quite simple. The real part of the load is nearly constant power and not affected by tap changing. The reactive part of the load may have relatively low voltage sensitivity. The shunt capacitor compensation, however, has a voltage-squared reactive power sensitivity. Thus, the main effect of tap changing is to support the capacitor output.

Example 4-7. An industrial load has motors that are shunt compensated. Consider the real part of the load to be voltage insensitive. Determine graphically and analytically the voltage sensitivity (at nominal voltage) of the net reactive load assuming 80% shunt reactive compensation. The uncompensated voltage sensitivity is $\Delta Q/\Delta V = 1.0$. What is the effect of tap changing?

Solution: Figure 4-32 shows the graphical solution. We see that the voltage sensitivity of the capacitor dominates and therefore the net reactive load will increase for voltage sag. Voltage regulation by tap changing will reduce net reactive load and improve voltage stability.

Analytically:

$$Q_{net} = Q_{load} + Q_{cap} = 1V^1 - 0.8V^2$$

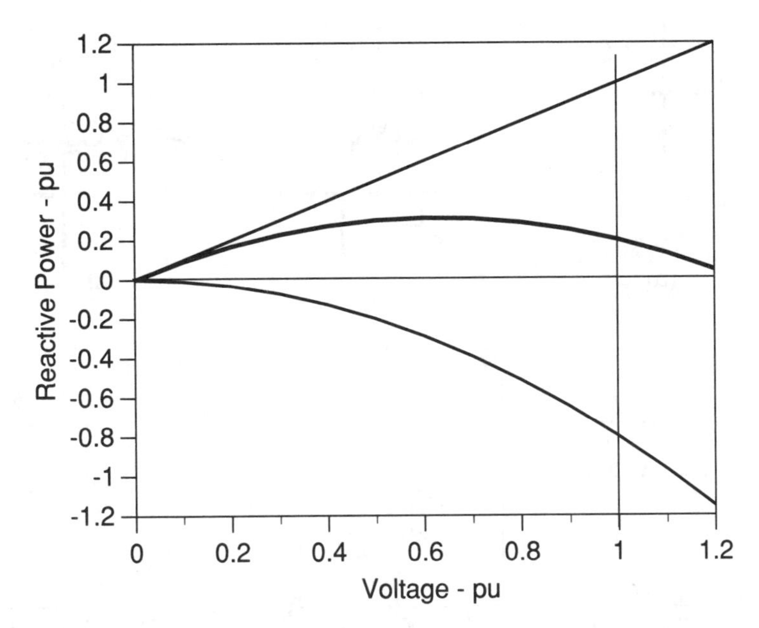

Fig. 4-32. Example 4-7: Graphical solution of net reactive voltage sensitivity of shunt compensated load.

$$\frac{dQ_{\text{net}}}{dV} = 1 - (0.8)\,(2)\,V = -0.6 \text{ at } V = 1 \text{ pu}$$

Note that the result depends on the voltage sensitivity of the reactive part on the load. As described above (Table 4-1 and Figure 4-11), voltage sensitivity depends on factors such as loading and is quite variable. Often, the reactive voltage sensitivity of the load may be higher than 1.0, resulting in less benefit from tap changing.

The support of capacitor bank output from voltage regulation should be considered in the placement of capacitor banks. Whenever possible, the capacitors should be on the regulated side of tap changers. In power flow simulation, capacitors should be represented on the proper bus. Figure 4-32 shows the effects of capacitor bank location on system performance and on power flow simulation [42]. Figure 4-33a shows the typical modeling of loads as constant power at a high voltage bus. Figure 4-33b shows the situation when capacitor banks are located on the regulated bus; constant real and reactive power is held until the LTC transformer reaches its boost limit. Figure 4-33c shows the effect if capacitor banks are either physically located on the high voltage bus or erroneously represented on the high side bus in simulation; the net reactive power load now increases as voltage falls because the capacitor voltage is not regulated.

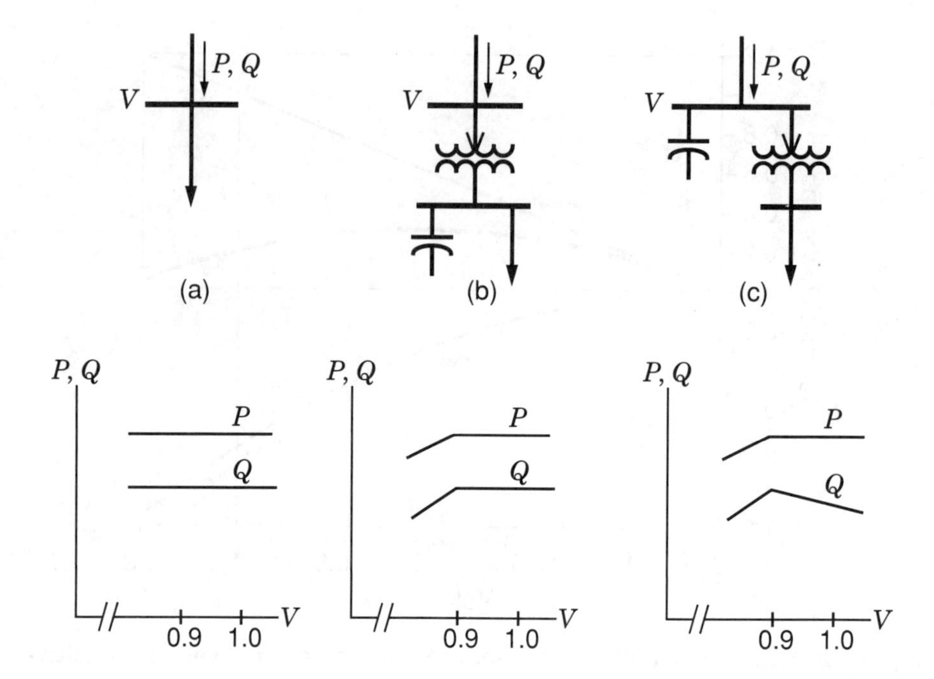

Fig. 4-33. Effect of capacitor location on net load characteristic.

Field test result. On 12 January 1989, the Bonneville Power Administration tested load response at Port Angeles on the Olympic Peninsula in Washington state. Figure 4-34 shows response of a radial 69-kV subtransmission line to shunt capacitor bank switching at the Port Angeles substation. The capacitor bank de-energization dropped the source voltage 4.5% during the twenty minute test period. The radial load is primarily residential with substantial electric heating. There are five bulk power delivery substations along the line, all with LTC transformers or voltage regulators (69/12.5-kV and 69/4-kV).

Immediately following the voltage drop, active and reactive load dropped about 7.75% and 29.3% respectively. This corresponds to voltage sensitivities $\Delta P/\Delta V = 1.73$ pu/pu and $\Delta Q/\Delta V = 6.5$ pu/pu. The high reactive voltage sensitivity could be mainly due to partial saturation of feeder transformers at normal voltage. However, the reactive power sensitivity calculation is not very reliable at high power factors.

The figure shows that voltage regulation (tap changing) has largely restored the active part of the load within about two minutes. The small restoration of reactive power is mainly due to increased losses (I^2X) caused by the active power restoration.

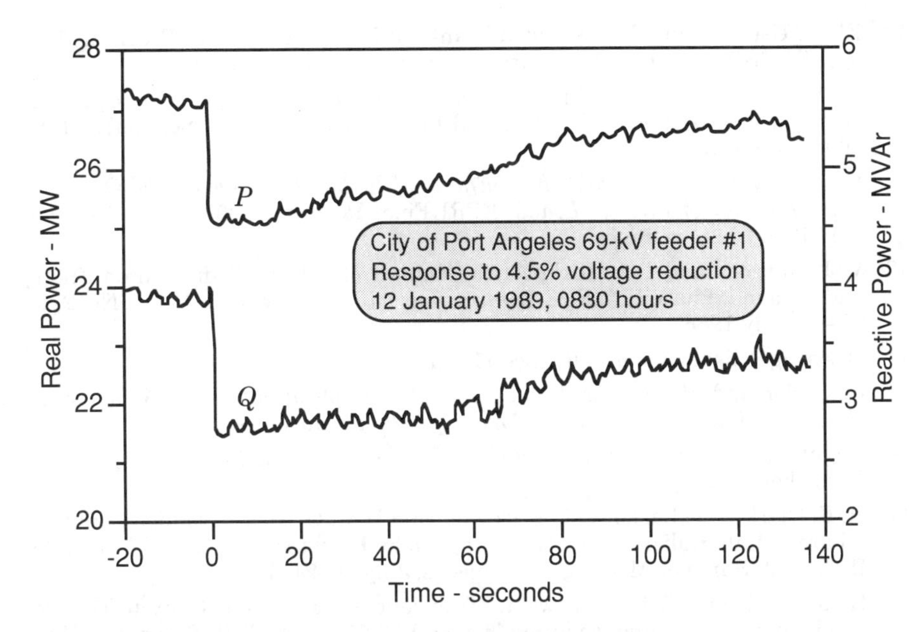

Fig. 4-34. Response of Port Angeles 69-kV subtransmission line to voltage drop. Initial load was 27.3 MW and 4.0 MVAr.

References

1. T. Gönen, *Electric Power Distribution System Engineering*, McGraw-Hill, New York, 1986.
2. University of Texas at Arlington, *Determining Load Characteristics for Transient Performances*, EPRI Final Report EL-848, May 1979, three volumes.
3. General Electric Company, *Load Modeling for Power Flow and Transient Stability Computer Studies*, EPRI Final Report EL-5003, January 1987. (The four volumes; describes LOADSYN computer program.)
4. W. W. Price, K. A. Wirgau, A. Murdoch, J. V. Mitsche, E. Vaahedi, and M. A. El-Kady, "Load Modeling for Power Flow and Transient Stability Computer Studies," *IEEE Transactions on Power Systems*, Vol. 3, No. 1, pp. 180–187, February 1988.
5. F. Nozari, M. D. Kankam, and W. W. Price, "Aggregation of Induction Motors for Transient Stability Load Modeling," *IEEE Transactions on Power Systems*, Vol. 2, No. 4, pp. 1096–1103, November 1987.
6. M. K. Pal, discussion of "An Investigation of Voltage Instability Problems," by N. Yorino et al., *IEEE Transactions on Power Systems*, Vol. 7, No. 2, pp. 600–611, May 1992.
7. D. S. Brereton, D. G. Lewis, and C. C. Young, "Representation of Induction-Motor Loads During Power-System Stability Studies," *Transactions AIEE*, Vol. 76, Part III, pp. 451–461, August 1957.
8. Arizona State University, *Midterm Simulation of Electric Power Systems*, EPRI Final Report, EL-596, June 1979.
9. P. Kundur, *Power System Stability and Control*, McGraw-Hill, 1993.

10. EPRI, User's Manual—Extended Transient/Midterm Stability Program Package (ETMSP Version 3.0), prepared by Ontario Hydro, June 1992.
11. University of Texas at Arlington, *Effects of Reduced Voltage on the Operation and Efficiency of Electric Loads*, EPRI Final Report EL-2036, September 1981. (Two volumes.)
12. University of Texas at Arlington, *Effects of Reduced Voltage on the Operation and Efficiency of Electric Loads*, EPRI Final Report EL-3591 June 1984 and July 1985 (three volumes).
13. V. J. Warnock and T. L. Kirkpatrick, "Impact of Voltage Reduction on Energy and Demand: Phase II," *IEEE Transactions on Power Systems*, Vol. 3, No. 2, pp. 92–97, May 1986.
14. N. Savage, discussion of reference 11, Ibid.
15. C. Crider and M. Hauser, "Real Time T&D Applications at Virginia Power," *IEEE Computer Applications in Power*, pp. 25–29, July 1990.
16. "Lighting the Commercial World," *EPRI Journal*, Vol. 14, No. 8, pp. 4–15, December 1989.
17. H. K. Clark, T. F. Laskowski, A. Wey Fo, and D. C. O. Alves, "Voltage Control in a Large Industrialized Load Area Supplied by Remote Generation," A 78 558-9, IEEE/PES Summer Meeting, Los Angeles, July 16–21, 1978.
18. H. K. Clark and T. F. Laskowski, "Transient Stability Sensitivity to Detailed Load Models: a Parametric Study," paper A 78 559-7, IEEE/PES Summer Meeting, Los Angeles, July 16–21, 1978.
19. S. Nadel, M. Shepard, S. Greenberg, G. Katz, A. T. de Almeida, *Energy-Efficient Motor Systems: A Handbook on Technology, Programs, and Policy Opportunities*, American Council for an Energy-Efficient Economy, Washington, D.C., 1991.
20. A. Domijan, Jr., O. Hancock, and C. Maytrott, "A Study and Evaluation of Power Electronic Based Adjustable Speed Motor Drives for Air Conditioners and Heat Pumps with an Example Utility Case Study of the Florida Power and Light Company," *IEEE Transactions on Energy Conversion*, Vol. 7, No. 3, pp. 396–404, September 1992.
21. B. R. Williams, W. R. Schmus, and D. C. Dawson, "Transmission Voltage Recovery Delayed by Stalled Air-Conditioner Compressors," *IEEE Transactions on Power Systems*, Vol. 7, No. 3, pp. 1173–1181, August 1992.
22. A. Kurita and T. Sakurai, "The Power System Failure on July 23, 1987 in Tokyo," *Proceedings of the 27th IEEE Conference on Decision and Control*, Austin, Texas, pp. 2093–2097, December 1988.
23. R. J. Frowd, R. Podmore, and M. Waldron, "Synthesis of Dynamic Load Models for Stability Studies," *IEEE Transactions on Power Apparatus and Systems*, Vol. PAS-101, No. 1, pp. 127–135, January 1982.
24. Y. Sekine and H. Ohtsuki, "Cascaded Voltage Collapse," *IEEE Transactions on Power Systems*, Vol. 5, No. 1, pp. 250–256, February 1990.
25. R. Betancourt and M. S. Lin, *A Study of the Effect of Transient Voltage Drop on Power System Stability*, San Diego State University, January 1986. (Includes an extensive list of references related to motor controllers.)
26. H. K. Clark, "Load Characteristics," 1990 WSCC Stability Seminar, April 1990.
27. IEEE Committee Report, "Load Representation for Dynamic Performance Analysis," paper 92 WM 126-3 PWRS, IEEE/PES 1992 Winter Meeting.

28. ANSI/IEEE Std 446-1987, *IEEE Recommended Practice for Emergency and Standby Power Systems for Industrial and Commercial Applications* (IEEE Orange Book) 1987.
29. J. Lamoree, "How Utility Faults Impact Sensitive Customer Loads," *Electrical World*, pp. 60–63, April 1992.
30. Klaus-Martin Graf, "Dynamic Simulation of Voltage Collapse Processes in EHV Power Systems," *Proceedings: Bulk Power System Voltage Phenomena—Voltage Stability and Security*, EPRI EL-6183, Section 6.3, January 1989.
31. D. Karlsson, K. Linden, I. Segerqvist, and B. Stenborg, "Temporary Load-Voltage Characteristics for Voltage Stability Studies - Field and Laboratory Measurements," *CIGRÉ*, paper 38-204, 1992.
32. D. Karlsson, *Voltage Stability Simulations Using Detailed Models Based on Field Measurements*, Technical Report no. 230, Chambers University of Technology, Göteborg, Sweden, June 1992.
33. H. Kirkham and R. Das, "Effects of Voltage Control in Utility Interactive Dispersed Storage and Generation Systems," *IEEE Transactions on Power Apparatus and Systems*, Vol. PAS-103, No. 8, pp. 2277–2282, August 1984.
34. T. J. E. Miller, editor, *Reactive Power Control in Electric Systems*, John Wiley & Sons, New York, 1982.
35. W. K. Wong, D. L. Osborn, and J. L. McAvoy, "Application of Compact Static Var Compensators to Distribution Systems," *IEEE Transactions on Power Delivery*, Vol. 5, No. 2, pp. 1113–1120, April 1990.
36. E. Larsen, N. Miller, S. Nilsson, and S. Lindgren, "Benefits of GTO-Based Compensation Systems for Electric Utility Application," *IEEE Transactions on Power Delivery*, Vol. 7, No. 4, pp. 2056–2064, October 1992.
37. Westinghouse Electric Corporation, *Electric Utility Engineering Reference Book—Distribution Systems*, 1965.
38. S. Elvin, "Using Series Capacitors for Distribution Networks," *Power Technology International 1991*, pp. 165–166.
39. J. W. Butler and C. Concordia, "Analysis of Series Capacitor Application Problems," *Electrical Engineering* (AIEE transactions), Vol. 56, No. 8, pp. 975–988, August 1937.
40. L. Morgan, J. M. Barcus, and S. Ihara, "Distribution Series Capacitor with High-Energy Varistor Protection," paper 92 SM 508-2 PWRD, IEEE/PES 1992 Summer Meeting.
41. M. S. Calovic, "Modeling and Analysis of Under-Load Tap Changing Transformer Control Systems," *IEEE Transactions on Power Apparatus and Systems*, Vol. PAS-103, No. 7, pp. 1909–1915, July 1984.
42. H. K. Clark, "Voltage Control and Reactive Supply Problems," 1988 WSCC Stability Seminar, April 1988.
43. W. R. Lachs and D. Sutanto, "Control Measures for Improving Power System Reliability," CIGRÉ paper 3A-01, *Symposium on Electric Power System Reliability*, Montréal, 16–18 September 1991.
44. W. J. Smolinski, "Equivalent Circuit Analysis of Power System Reactive Power and Voltage Control Problems," *IEEE Transactions on Power Apparatus and Systems*, Vol. PAS-100, No. 2, pp. 837–842, February 1981.

5

Generation Characteristics

Nothing in life is to be feared: it is to be understood.
Madame Marie Curie

The synchronous generator with its controls is one of the most complex devices in a power system. We have described longer-term voltage instability as usually being caused by two primary factors: load restoration by tap changing and generator current limiting—our interest is now the latter.

Power system disturbances leading to voltage instability often involve generation-load imbalances. This causes redistribution of power flow and reactive losses. We must understand how power plants respond to these upsets.

First we concentrate on generators and excitation systems, and then on prime movers and energy supply systems.

5.1 Generator Reactive Power Capability

The overexcited reactive power supply capability of synchronous generators is critical in preventing voltage instability.

Generator capability curves and V curves. For the slower forms of voltage stability, the steady-state generator P–Q capability curves and V curves are our starting point.

With the real power loading fixed, the allowable reactive power is limited by either armature (stator) or field (rotor) winding heating. As developed in several textbooks [1–3], the armature and field winding heating limits can be represented by two curves on a real power/reactive power plot. The intersection of the two curves represents the generator rated power factor. Figure 5-1 shows development of capability curves for two generators: a 0.8 power factor generator with relatively low synchronous reactance and a 0.95 power factor generator with high synchronous reactance. The latter is typical of recent generator design and specification.

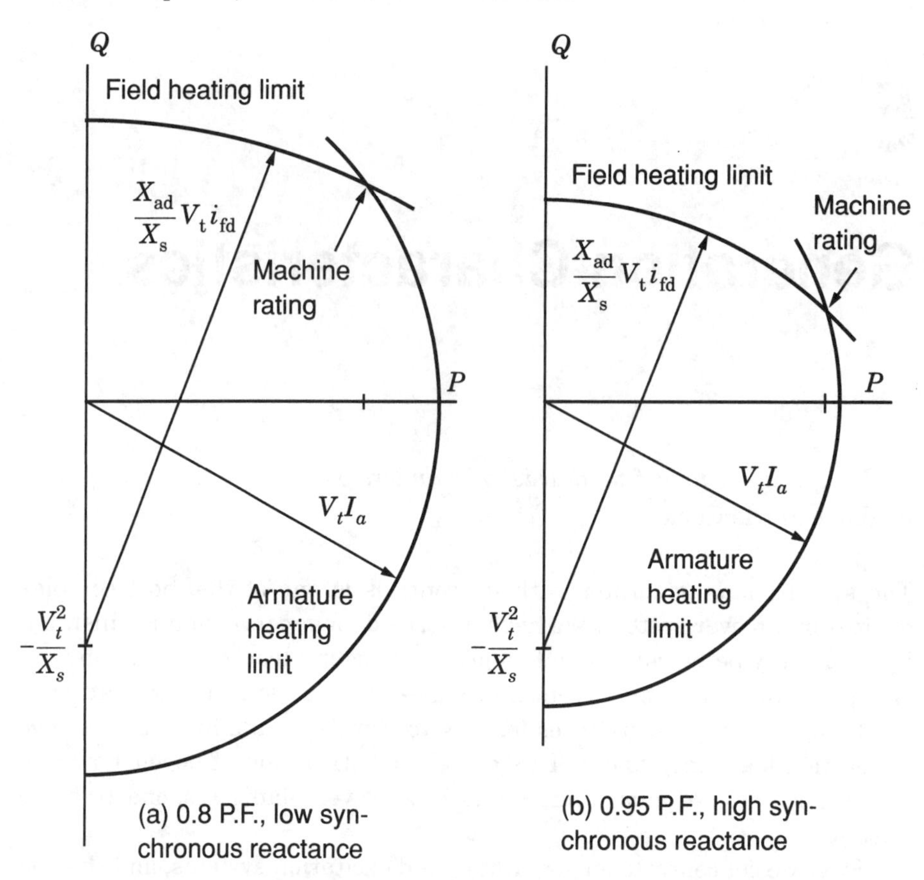

Fig. 5-1. Development of generator capability curve [2].

Figure 5-2 shows the V curves for a 410-MVA round-rotor generator. For the several values of cooling system pressure shown, the right-hand side limit is due to field heating. Figure 5-3 shows the generator capability curves corresponding to the V curves of Figure 5-2. The generator capability is greatly affected by the cooling system—as shown by the curves for different hydrogen cooling pressures. At lagging power factors lower than 0.9, the limiting factor is field winding heating. Between rated power factor (0.9 lagging) and 0.95 leading, the limiting factor is armature current. For lower leading power factor, the limiting factor is armature core end heating [3,4] or system stability.

In the portion of a capability curve where armature current is limiting, the power capability obviously varies with terminal voltage. The field winding curve also varies with terminal voltage [1–3,5].

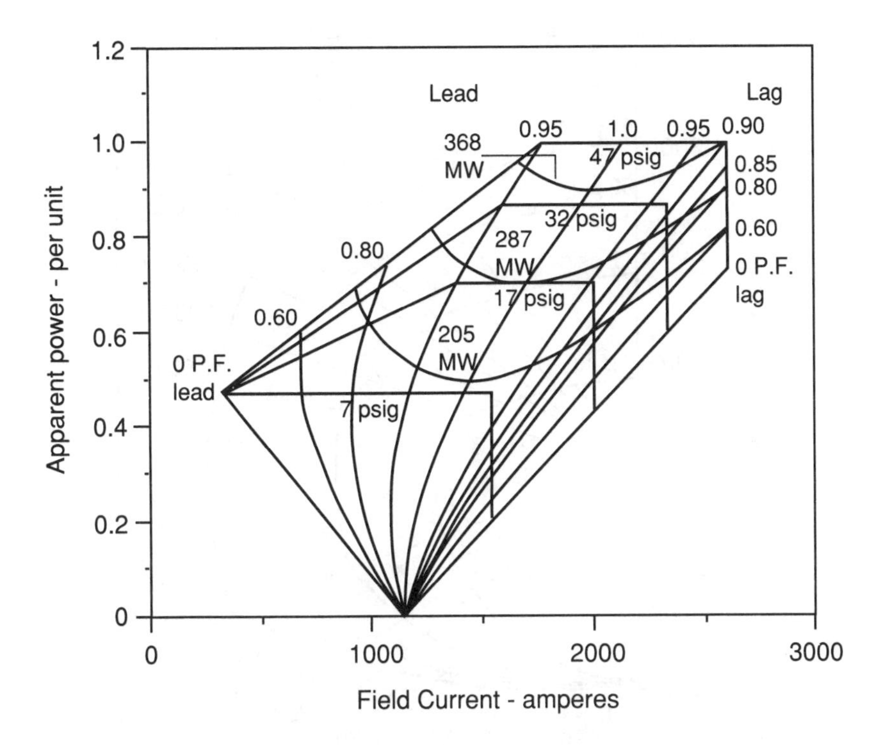

Fig. 5-2. Generator V curves. Generator nameplate data: 410-MVA, 3600 RPM, 22-kV, 0.9 power factor, 0.58 SCR, 47 psig, hydrogen pressure, 500-volt excitation.

Figure 5-4, adapted from Lachs [6], shows the overexcited portion of a generator capability curve. The figure shows a typical situatior. where the turbine is sized to match the generator real power at rated power factor. Also shown is the effect of reduced terminal voltage. We see that reactive power is normally limited by field heating. Armature current is limiting only if terminal voltage can no longer be controlled—which might be the case in a voltage emergency. It is apparent that armature current limitation depends on the maximum turbine power.

Example 5-1. A company normally purchases generators with 0.95 power factor. Turbine rating is specified to match the real power at rated power factor. Keeping the turbine rating unchanged, calculate the generator rating if 0.8 power factor is specified. Calculate the additional reactive power capability at full load.

Solution: Figure 5-1 shows the two generator types. Let the turbine power rating be one per unit. Since P.F. $= \cos\phi = (P/S)$, $S = 1.0525$ for

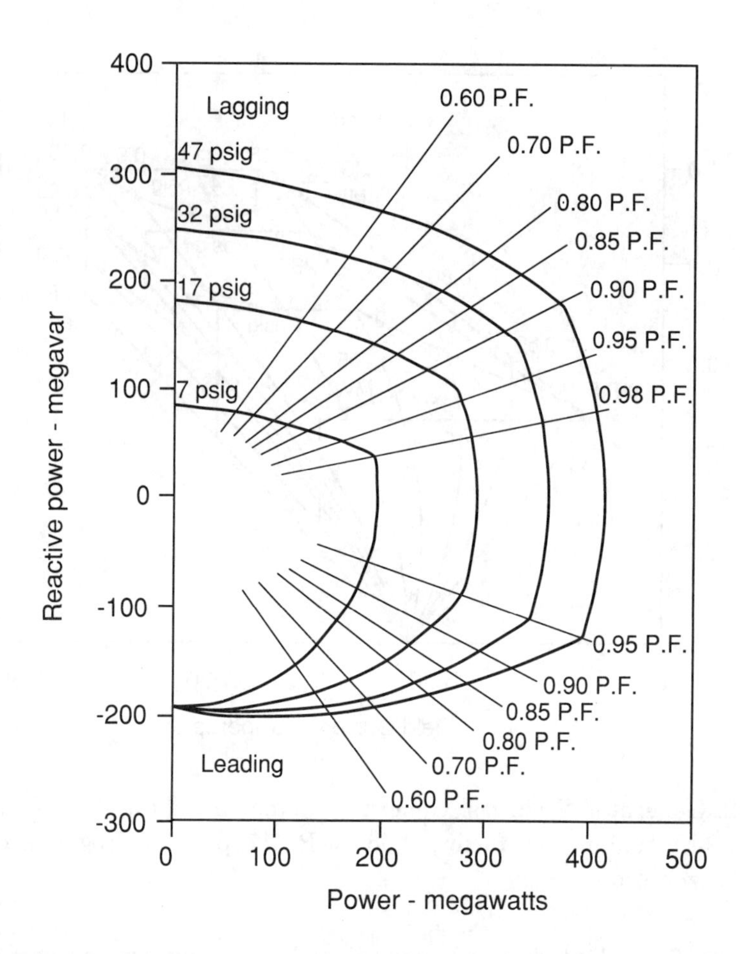

Fig. 5-3. Generator capability curve for the 410-MVA generator whose V curves are shown on Figure 5-2.

0.95 power factor and $S = 1.25$ for 0.8 power factor. The increase in MVA rating is 18.75%, but the increase in capital cost will be somewhat less than 18.75%. Assuming normal operation near unity power factor, the larger machine with more armature copper will have lower losses; this will reduce the life-cycle cost increase.

Using $Q = P \tan \phi$, the reactive power capability at rated conditions is 0.33 per unit for the 0.95 power factor machine. The reactive capability is 0.75 per unit for the 0.8 power factor machine, or 227% of the smaller machine. The additional controllable reactive power capability could be crucial during a voltage emergency.

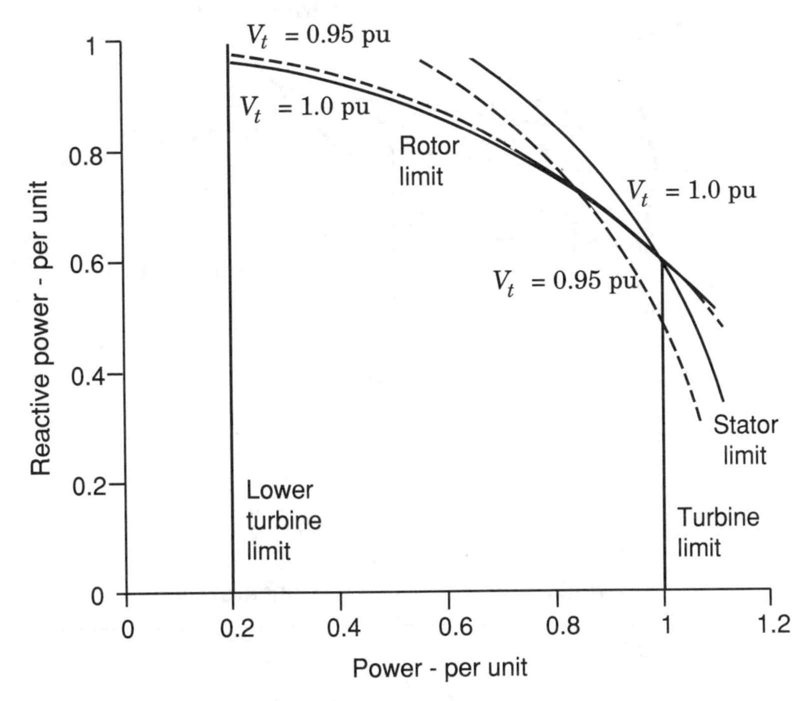

Fig. 5-4. Overexcited portion of generator capability curve showing effect of reduced terminal voltage [6]. © 1979 IEEE.

An effective method to improve voltage stability is to apply shunt capacitor banks so nearby generators can operate near unity power factor with substantial spinning reactive power reserve that can be rapidly activated to prevent voltage instability.

Generator Q–V curves. Although it is common to show performance of generators by P–Q capability curves, the important characteristics for voltage problems can be better shown by Q–V curves.* Such diagrams are constructed from a series of capability curves for different network (generator high-side) voltages, V. The diagram is for the generator plus its step-up transformer at a single value of active power. From a system viewpoint, the network voltage is most important.

Figure 5-5 shows a Q–V diagram. The sign convention for reactive power is the same as for a reactor, capacitor, or SVC. Curves for constant

*The following is due to Mr. Torben K. Østrup, Elkraft Power Company, Denmark [7].

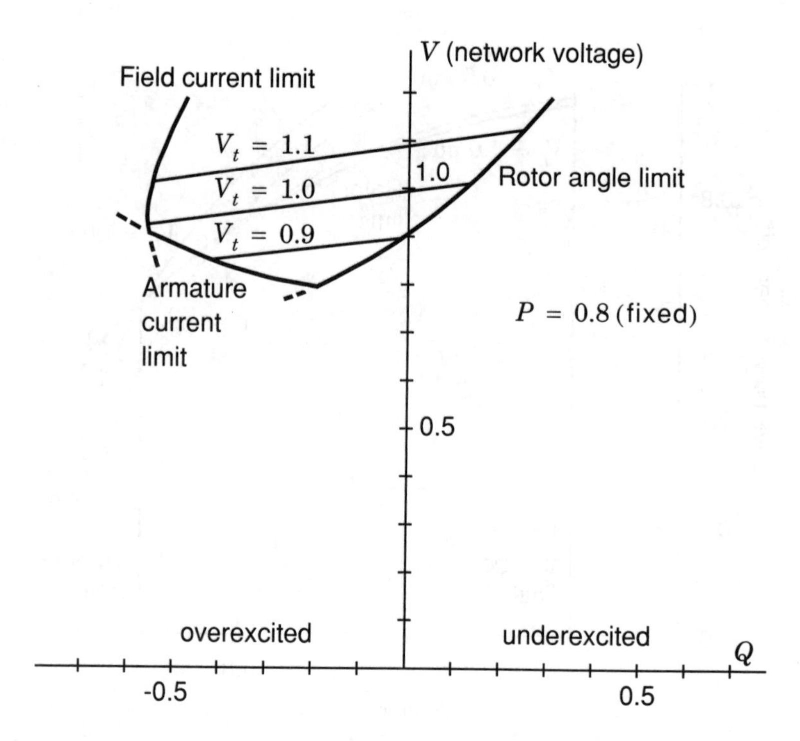

Fig. 5-5. Generator Q–V diagram for constant value of active power. In many cases the rotor angle limit, which is not of concern for voltage stability, may be less restrictive.

terminal voltage, V_t, are shown. If the reference value for the terminal voltage, V_t, and the network voltage, V, are given, the reactive power of the generator can be found from the figure.

When the generator is regulating voltage, the curves for constant V_t are very flat. This indicates that a large change in reactive power output is obtained for a small change in network voltage. Generators keep the network at near constant voltage, thus preventing voltage collapse. The characteristic is similar to a SVC in its active control range.

If the network voltage becomes sufficiently low, either the field current limit or the armature current limit of the generator is reached. This drastically changes the characteristic of the generator. The slope of the rotor current limit curve is nearly vertical (nearly constant reactive power output), meaning that voltage support from the generator is lost when this limit is reached.

Much worse conditions result if the armature current limit is reached. With the armature current limit enforced, the reactive power from the generator decreases fast if the network voltage is further lowered. Voltage instability is likely. Note also that limits on acceptable terminal voltage may be violated and power plant auxiliary equipment problems may result.

Performance under normal and current limiting conditions can be compared with SVCs and shunt capacitors. For field current limiting, the characteristic is less severe than for SVCs at maximum capacitive boost. Armature current limiting is more severe. Synchronous condenser performance is similar to generators.

The generator *Q–V* curve can be rotated ninety degrees clockwise and superimposed on a system *V–Q* curve. The curve should first be replotted so reactive power on the network side is shown.

Generator capability polyhedron. We can view generator capability as a function of network voltage in three dimensions by the capability polyhedron introduced by Professor Calvaer [8]. Again, network voltage is plotted as this is most important for system performance. Figure 5-6 shows a capability polyhedron for a 1330-MVA unit connected to a 400-kV grid.

Synchronous machine steady-state models. In conventional power flow programs, generators are represented as "PV buses" with real power specified and reactive power output between minimum and maximum limits (see Appendix B). At a reactive power limit, a PV bus is changed to a PQ bus. The maximum value, Q_{max}, is usually the reactive power at rated conditions. This results in a rectangular generator capability curve. If generators are operated below rated power, the additional available reactive power should be recognized. Many relatively new generators have high rated power factor and high synchronous reactance (low short-circuit ratio). As Figure 5-1 shows, this makes the difference between a rectangular capability curve and the actual capability curve more important [5].

Referring back to Figures 5-4 and 5-5, an approximate model could simply prevent armature current overload whenever reactive power limiting is in effect at a generator (whenever a PV bus is changed to a PQ bus). In other words, Q_{max} is a function of terminal voltage. Depending on actual generator control and protection, time frame, and operator actions, this may or may not be an adequate model.

Section 5.2 describes generator control and protection which must be represented in detailed simulations. Advanced power flow programs for voltage stability analysis enforce field current limits by direct computation of field current. The computation is similar to generator initialization in transient stability simulation [3], and includes representation of generator saturation. Rather than by a PQ bus, a generator in field current limiting is modeled by a voltage behind saturated synchronous reactance; alternatively, a two-axis generator model may be used with fixed field voltage. With drop in terminal voltage, the armature current should be monitored for overload. For normal conditions, advanced programs may also replace the PV bus model with a model that includes the voltage regulator droop effect on terminal voltage (voltage regulators are finite gain, proportional controllers, and terminal voltage will vary with loading.) Field current limiter settings, and armature current alarm and protection limits are required for accurate simulation

Operating practices. In some situations, voltage stability can be improved by reducing real power on certain generators so that more reactive capability is available. If, during a severe disturbance, a generator is feeding a highly overloaded line, stability can be improved by reducing real loading and rescheduling the generation at power plants serving lightly loaded lines into the load area.

It's vital that the reactive power expected from a generator's capability curve actually be available for emergencies. Some utilities have power plant test programs to verify that maximum and minimum reactive power limits can be reached [9]. For operations, it's important to know the reactive capability, as affected by cooling system status.

During normal conditions, many generators operate at high power factor. Power plant operators should be trained in the basics of voltage stability and instructed not to reduce generator reactive output during rare system emergencies when high levels of reactive power output are needed.

Generator armature and field overload capabilities. The armature and field overload capabilities given by ANSI C50.13-1977 are listed below:

Armature				
time (seconds)	10	30	60	120
armature current (percent)	226	154	130	116
Field				
time (seconds)	10	30	60	120
field voltage (percent)	208	146	125	112

The time-overload capabilities of armature and field windings are similar. Note that, at most, overloads should last only two minutes. Excitation could be controlled to allow a high overload for less than a minute or a mild overload for up to two minutes.

Overload capability depends on the status of cooling systems. Operator response to overtemperature alarms could lead to limiting of generator output.

Exciter equipment (excitation transformer or exciter alternator) overload characteristics must also be considered.

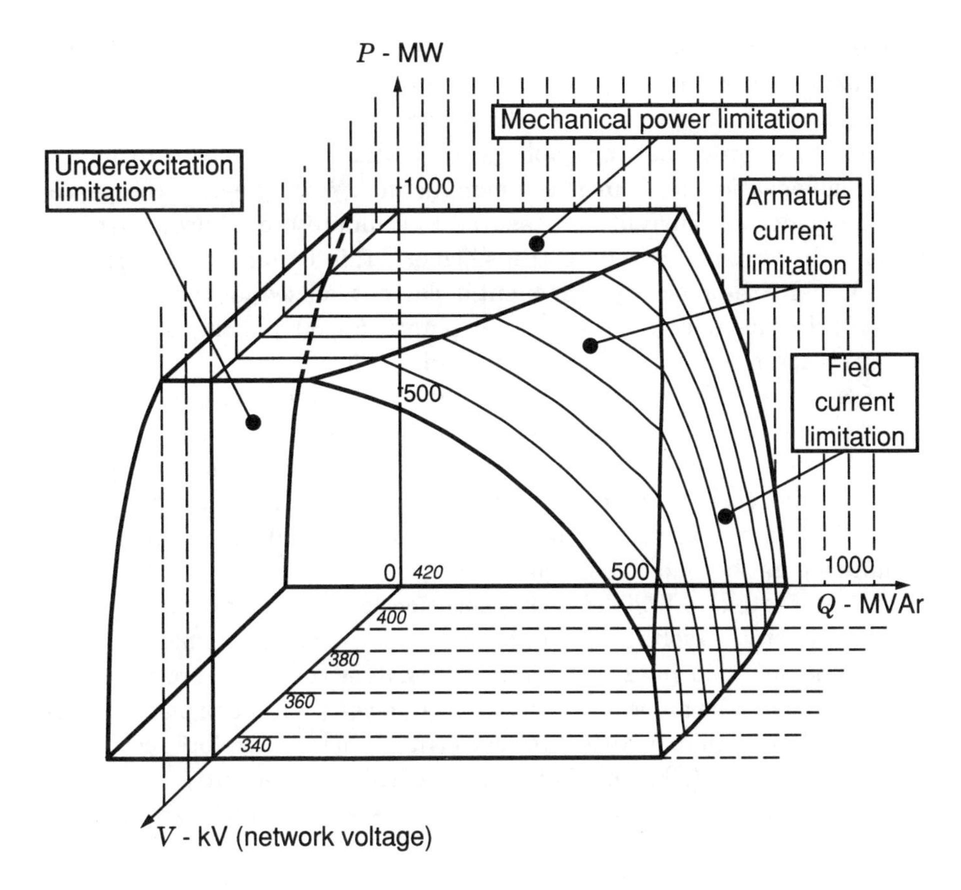

Fig. 5-6. Capability polyhedron for a 1330-MVA generator. Source: F. Van de Meulebroeke, Laborelect, Belgium [7].

5.2 Generator Control and Protection

Excitation control is by automatic voltage regulators (AVRs). Generator terminal voltage is measured and compared to a desired or reference setpoint. The error signal controls the exciter output which is the main generator field voltage. During steady-state operation, terminal voltage is closely regulated by the high gain feedback control system. Typically the terminal voltage error is ±0.5% from no load to full load (1% slope or droop). There is also a small droop due to generator reactive power changes. The small voltage regulating droop can be compared to similar small droops (1–5%) of static var compensators. A difference is that a SVC regulates the high voltage or extra-high voltage network bus voltage, while a basic AVR controls generator terminal voltage.

Network voltage control. Voltage stability can be greatly improved if generators regulate network voltage (step-up transformer high-side voltage). Network voltage should be held as high as possible. There are several methods to improve network voltage regulation; line drop (transformer drop) compensation is the most common method. With line-drop compensation, the generator current is measured to compute the *IZ* drop part way through the generator step-up transformer—this becomes the regulated point. Usually the reactive component of current is used for the compensation [10]. In multiple generator power plants, 50–80% of the transformer leakage reactance is typically compensated.

Line-drop compensation is more difficult when several units are connected to the same low-side bus—a typical situation in hydroelectric plants consisting of many units. When two or more generators regulate the same point, reactive current compensating circuits or other methods are needed to equalize reactive power sharing and control effort [10,11]. It's possible to regulate a point beyond the network bus.

An alternative to line-drop compensation is a secondary outer control loop to adjust the voltage regulator setpoint in order to maintain the desired network side voltage. To minimize adverse control interactions, the outer loop should be an order of magnitude slower than the primary terminal voltage regulation. Network voltage control should therefore respond in around ten seconds following a disturbance. This is sufficiently fast for the slower forms of voltage instability, i.e., it's faster than the restoration of load by tap changing.

Tokyo Electric Power Company has developed a special controller for fast regulation of network voltage [12]. Using programmable logic controllers, Manitoba Hydro has developed a 230-kV voltage controller for nine synchronous condensers at the Dorsey HVDC inverter station [13,14]. French, Italian, and Belgian utilities are developing a "secondary voltage control" centralized method for controlling network voltages and generator reactive output [15,16,17]; these techniques are slower and more expensive than local methods.

Still another possibility for better network voltage control is load tap changing of generator transformers. Presumably for economic and reliability reasons, LTC generator transformers are not used in North America. Many overseas utilities, however, use this method. LTC generator transformers or LTC auxiliary equipment transformers prevent constraints on terminal voltage magnitude from limiting high side voltage control. For fixed tap transformers, tap settings of main and auxiliary transformers should be optimized taking into account voltage stability performance.

For voltage stability, optimal performance is when, following a severe disturbance, all generators reach field current limits at about the same time. Appropriate settings can generally be determined off line.

Excitation control for transient voltage stability. Fast excitation control is important for transient stability, including transient voltage stability problems associated with induction motor loads or HVDC links. The speed of generator and excitation system dynamics are similar to the speed of the load dynamics. Static exciters and other high initial response exciters will improve stability. High exciter ceiling voltages allowing momentary high field overload are important. P. Kundur covers these topics in the companion book on rotor angle stability [3].

Generator field control and protection. Besides the basic voltage regulating function, AVRs have several other functions. Our interest is in the circuits that limit field current.

Most automatic voltage regulator include an overexcitation limiter or "maximum excitation limiter" which detects high field voltage or current and acts through the regulator to return excitation to a preset value after an adjustable time delay. Brushless exciters do not allow direct measurement of generator field current or voltage; either exciter field, or generator armature quantities must be used.

The overexcitation limiter has an inverse timing characteristic coordinated with the field thermal capability and field time-overcurrent relay. (Older equipment has a fixed time delay rather than an inverse characteristic.) Baldwin and McFadden describe the control and protection provided by one manufacturer [18]; Figure 5-7 shows the coordination.

Quoting from Baldwin and McFadden:

> The Maximum Excitation Limiter (MXL) acts through the regulator to return the level of excitation to a preset value after an adjustable time delay during which overexcitation is permitted. ...Since the limiter acts through the regulator, it cannot provide protection of the field for all operating modes, such as manual control, or in the event of malfunction of the regulator itself.
>
> The limiter is therefore supplemented by the overexcitation protection (OXP). This device with inverse timing also has a characteristic which matches the generator field capability. The OXP initiates sounding an alarm and running the base adjuster to a preset position, usually that corresponding to rated conditions for the generator. It also initiates regulator trip after the preset interval has passed. If the tripping of the regulator and repositioning of the base adjuster has not reduced the excitation, it will also initiate a

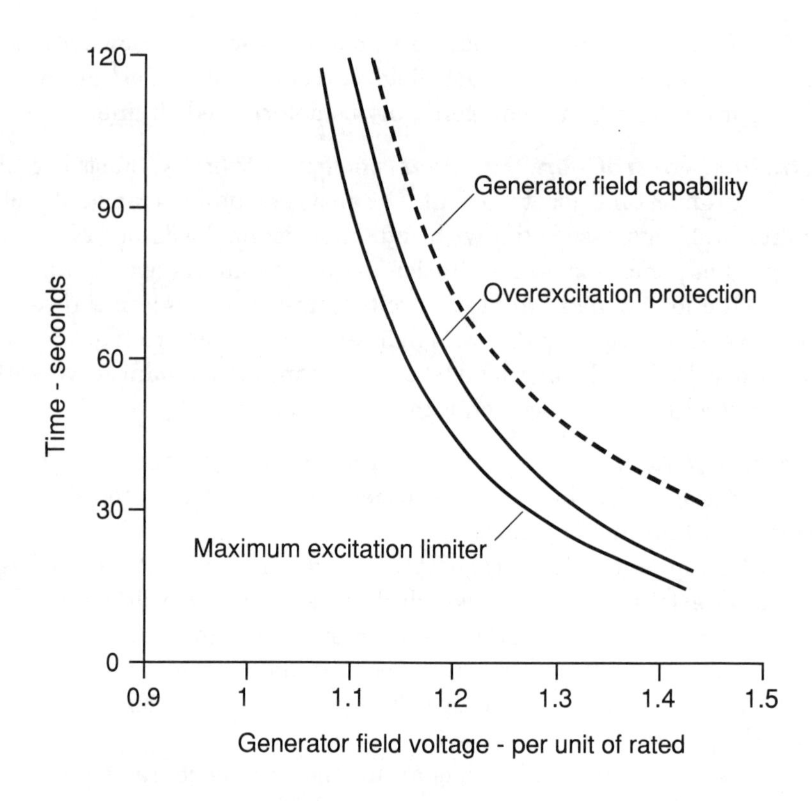

Fig. 5-7. Coordination of overexcitation limiting and protection. © 1981 IEEE. The generator field capability corresponds to ANSI C50.13-1972.

> unit trip after an additional fixed time delay, if the user so desires. © 1981 IEEE [18].

In other cases, an inverse-time overcurrent relay is used. In one type, relay operation recalibrates an AVR current limit circuit to rated field current; operators must manually reset the control circuit to return to normal voltage regulation. In another type, operation of the relay will trip the AVR to manual; depending on the base adjustment this may result in a large drop in excitation and reactive power. Either type may trip the generator if the overexcitation is not relieved. In older installations, the overcurrent relay may simply trip the generator.

There are a wide variety of overexcitation control and protection devices in service. Often the power plant must be visited to determine the exact equipment in service and the settings. Replacement of an old excitation system with a modern system employing a continuous field current limiting control loop may be highly desirable.

Overexcitation limiter modeling. Standard models for excitation limiters have not been developed. The companion book [3] provides a detailed model developed for a particular power plant. Figure 5-8 shows a typical model for modern equipment that would often suffice [7].

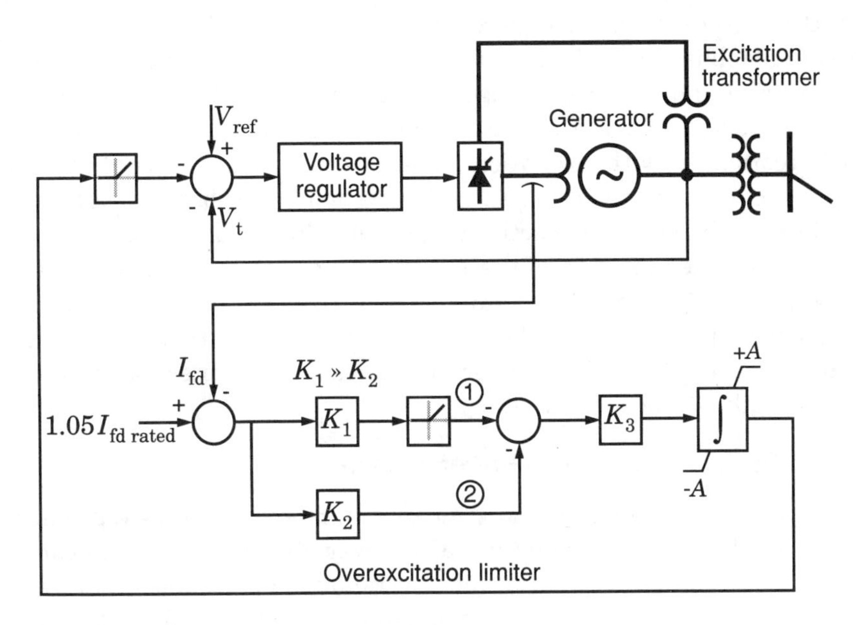

Fig. 5-8. Model of overexcitation limiter.

During normal conditions with field current less than the setpoint (typically 105% of rated field current), paths 1 and 2 both force the integrator to its lower limit (-*A*). The input to the voltage regulator summing junction is zero. If field current is above the setpoint, path 2 drives the integrator towards a positive value; once the integrator output becomes positive, the voltage regulator will start to reduce field current. For a step increase in field current above the setpoint, the time to the start of current limiting (time for integrator output to ramp to zero) will be:

$$t = \frac{A}{K_2 \bullet K_3 \bullet (I_{fd} - 1.05 I_{fd\ rated})}$$

The parameters can be adjusted so this time approximates the desired time as shown on Figure 5-7. Once field current is reduced to the setpoint, path 1 rapidly resets the integrator to -*A*.

Generator armature protection and system backup protection. An IEEE guide [19] describes generator protection in detail. Most generators are supplied with a number of resistance temperature detectors (RTDs) which are imbedded in the armature windings. The RTDs may be supplemented with time-overcurrent relays that are coordinated with the armature short-time overcurrent capability. The RTDs and overcurrent relays provide alarm signals, or trip the unit.

Distance relays or voltage-restrained or controlled overcurrent relays are often applied on generators for "system backup" protection [19]. Armature current and terminal voltage measurements drive the relays. Usually the time delays of these relays are one second or less—this means they don't coordinate with excitation limiting controls just described. These relays are meant to detect transmission system faults, but could operate on overload (high generator reactive output) and low terminal voltage.

Undesirable operation of these relays have contributed to several voltage collapses. Where voltage stability is a concern, the settings of these relays should be reviewed. Sensitive protection prone to operate on overload may not be necessary where redundant line protection relays and breaker failure relays are used.

5.3 System Response to Power Impacts

By power impact we mean load switching or, more importantly, generation loss. Loss of a large generator or entire power plant in a load area can be a severe disturbance threatening voltage stability. A particularly severe case in the Northeastern U.S. is loss of imports from HVDC links with Quebec. In other circumstances, generation, especially hydro generation, may be intentionally tripped to prevent voltage instability. We now describe how a system of many generators responds, over time, to a generation-load imbalance. This will result in terms such as "governor" and "AGC" post-disturbance power flow.

At any point in time, generation must match load plus losses. How lost generation is redistributed to remaining generation determines transmission line flows. If heavily loaded transmission lines become even more heavily loaded, reactive power losses increase to threaten voltage stability.

Time sequence of generation response. Immediately following loss of a generator, how is the lost power accounted for so that generation matches load plus losses? How is the power redistributed? Because some loads are voltage sensitive, there will be load reduction owing to voltage sags. Most of the lost generation, however, is made up by the remaining generators. For a fraction of a second, the replacement generation comes from stored magnetic energy. The generators near the lost generation pick up the most. Power redistribution is according to electrical closeness or "synchronizing coefficients." In transient stability simulation, the redistribution is the power flow at t = 0+. The generator internal voltage or flux has not yet decayed. Within a fraction of a second, however, "armature reaction" caused by the increased current will cause flux decay—which is countered by excitation control.

What happens next? Generators slow down and frequency starts to decay because of the generation loss. The inertia of the generator-turbine rotors limits the rate of frequency decay. The nearby machines deccelerate because of their overload, leading to lightly damped electromechanical oscillations between all generators. After a few seconds, the oscillations are damped out, and the power redistribution is by relative inertia values (stored mechanical energy). The larger generators pick up most of the lost power regardless of location. This is sometimes called "inertial" power flow. There is also some reduction in loads due to the reduced frequency.

During roughly the same time period, prime mover controls (governors*) respond to the decaying speed and open steam-turbine valves or hydraulic-turbine gates. Prime mover controls arrest the speed and frequency decay after 3–5 seconds. The inherent response of turbines and motors with speed-dependent loads help arrest the speed and frequency decay. Assuming oscillations have damped out, frequency is uniformly low throughout the interconnection. After 10–20 seconds, frequency has partially recovered and power redistribution is largely by prime mover control (stored thermal or hydraulic energy). All units in the interconnection with active governors will participate and again the larger generators pick up most of the lost power regardless of location. This is sometimes called "governor" power flow [21]. Governor power flow may represent the time frame of longer-term voltage stability.

Anderson and Fouad [22, Chapter 3] and Rudenberg [23] describe the mathematics of the fast phenomena just described.

Prime mover control does not restore frequency to schedule (60 Hz). Prime mover control provides proportional rather than integral control. Automatic Generation Control (AGC) is the mechanism to restore frequency and net interchange power to scheduled values by integral control. AGC works over tens of seconds and minutes. AGC may take 10–15 minutes for very large imbalances.† The lost power is now made up by generators in the interchange control area that experienced the loss of

---

*An IEEE standard [20] depreciates the use of *governor* "to advance the understanding that control systems need not be limited to rotating flyweights but include mechanical, hydraulic, and electronic components."

†The North American Electric Reliability Council (NERC) requires that generation lost be covered within ten minutes. Utilities are not always successful in meeting this requirement.

generation (stored energy in fuel supply). If the control area does not have enough reserve generation, interchange schedules must be altered and other generators cover the loss. This is sometimes called "AGC" power flow and is the condition several minutes after the outage. This is the condition normally represented in power flow outage cases where area interchange control models AGC. Extension of AGC power flow to an even longer time frame is "economic dispatch" power flow.

In summary, for power flow analysis of the slower forms of voltage stability, we are interested in either governor or AGC power flow. Sometimes we need to study both time frames. Power redistribution is by two completely different criteria. In governor power flow, generating plants throughout the synchronous interconnection participate. In AGC power flow, only selected generating units participate. For large generation-load imbalances, it may take many minutes for the transition from governor to AGC power flow. Appendix C provides guidance for power flow simulation.

Example 5-3. A load area with local generation is interconnected with a large system as shown in Figure 5-9a. One of the load area generators suddenly trips. Sketch the resulting response of the remaining local generator and the response of the system equivalent generator. Also sketch response of system frequency.

Solution: Figure 5-9 shows the responses for the power impact phenomena described above.

In some situations, intentional generator tripping improves voltage stability. To keep the reactive support, fast generation runback may, instead, be used. Hydro generation tripping is often used since hydro generation is easy to resynchronize and reload.

Three examples from the western North American interconnection follow.

1. Pacific Intertie system. The Pacific Intertie consists of a 3100-MW, ±500-kV, 1360 km HVDC line in parallel with a 3200 MW, 500-kV ac link (two or three parallel circuits per section). See Figure 5-10. Bipolar outage of the HVDC Intertie presents a very severe disturbance to the power system, and to the parallel Pacific AC Intertie in particular. For HVDC Intertie outages, Pacific Northwest hydro generation (up to 2700 megawatts) is tripped for three reasons: maintain transient stability, reduce overload on AC Intertie series capacitors, and maintain longer-term voltage stability in Northern California. Voltage instability on the AC Intertie following loss of the HVDC intertie would probably blackout the entire Southwestern U.S.

Voltage stability became a concern following a bipolar HVDC outage (1286 MW) on May 21, 1983. After about two minutes, voltage on the

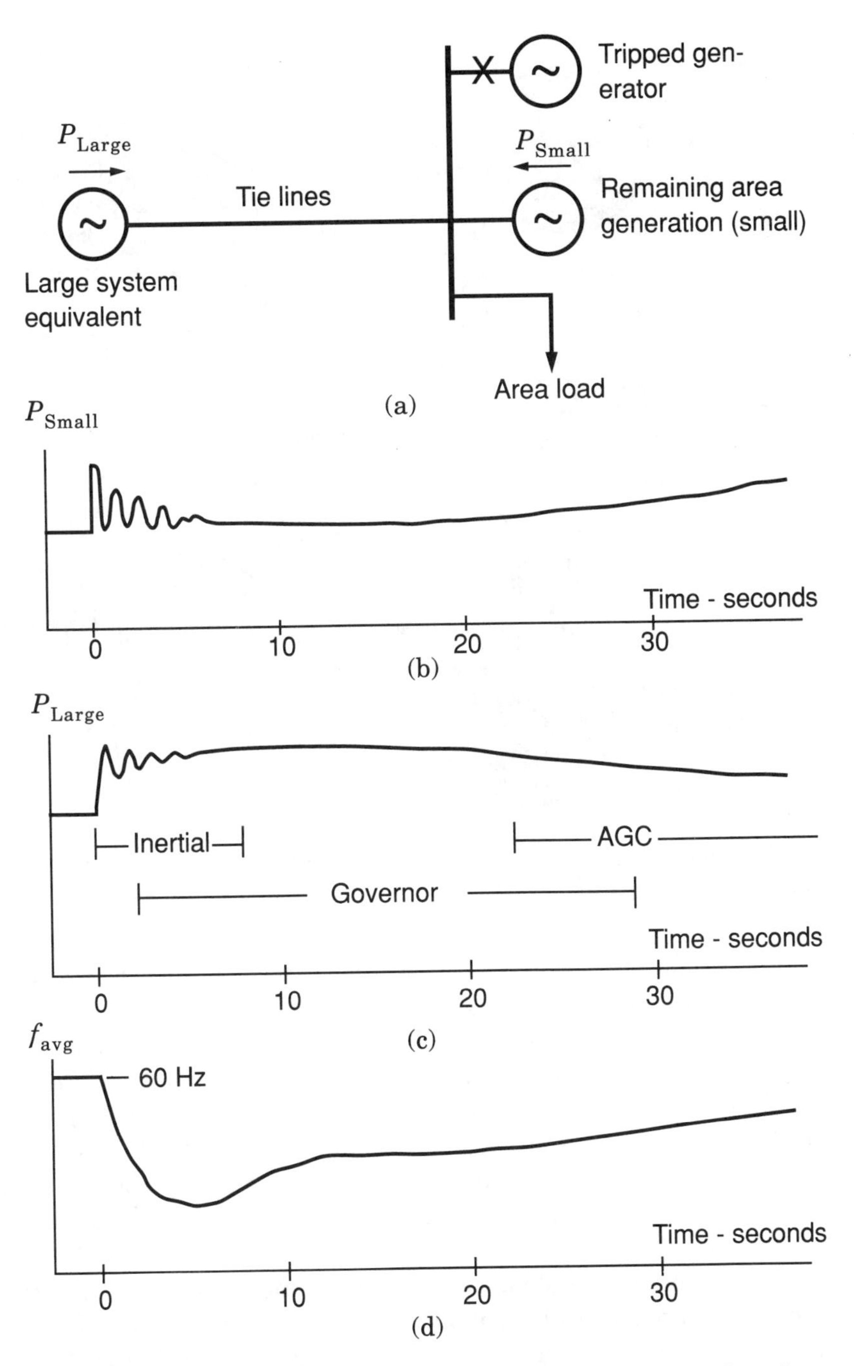

Fig. 5-9. Power system response to power impact. We assume that the control area with loss of generation has enough spinning reserve to cover the loss generation.

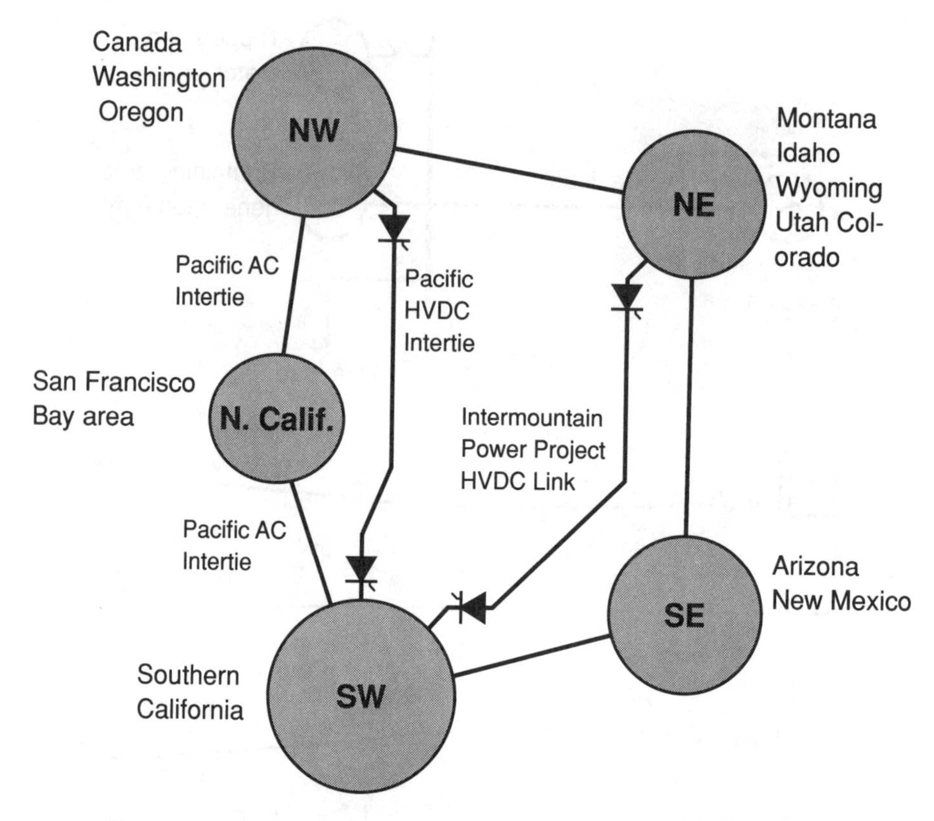

Fig. 5-10. Schematic of western North American interconnection and Pacific Intertie system.

500-kV intertie north of San Francisco decayed to 380 kV (0.76 pu). Voltage recovered after pump motors at aqueduct stations tripped.

In tripping Pacific Northwest hydro generation, redistribution of generation by prime mover control or AGC becomes critical. The more power picked up in the southern part of the system, the better the voltage stability. Ideally, AGC at the southern end should make up for the lost HVDC import within about two minutes. (This could possibly be done by special fast power changes at pumped hydro stations and at the Hoover Dam hydro plant.)

2. The British Columbia, Canada system includes remote generation sites on the Peace River and Columbia River. Outages of key 500-kV lines between the remote generation and the Vancouver load area threaten longer-term voltage stability [24]. B.C. Hydro is strongly connected to the U.S. system—by two 500-kV ac transmission lines plus lower voltage lines.

Following line outages, tripping remote generation on the appropriate river system would relieve heavily loaded lines with high reactive power losses. The generation tripped would largely come from the much larger U.S. system—first by governor power flow and then by AGC schedule changes. Power normally flows north to south, from the Vancouver area to the Seattle area; the generator tripping would cause the power to reverse, and the power supply from a strong source would maintain voltage stability. (This control is not implemented.)

3. As described in Chapter 7, the Puget Sound area in the Pacific Northwest has longer-term voltage stability problems associated with heavily loaded 500-kV transmission lines running east to west from Columbia River generation in Eastern Washington state to the Puget Sound area. Following 500-kV line outages, tripping (or run-back) of Columbia River generation would relieve heavily loaded lines and improve voltage stability, particularly if the lost power could be delivered on lightly loaded north/south lines into the Puget Sound area. This could involve, for example, fast increase of British Columbia hydro generation. (This control is not implemented.)

5.4 Power Plant Response

We have hinted about needs to rapidly increase generation to aid longer-term voltage stability. Often the fast power changes must be made within a few minutes. We are in a race with load restoration by tap changing and with generator current limiting. Recalling Figure 1-5, transmission line reactive power consumption increases nonlinearly with real power loading increases. We thus want to unload heavily loaded lines, bringing power into the load area on lightly loaded lines. Generation should be increased in areas where the increased generation does not cause reduction in generator reactive power capability at critical locations. Figure 5-11 shows the situation schematically.

Small generation changes by governors are quite fast—a few seconds. We are now interested in how fast large changes can be made by either conventional AGC, or by special generation controls used to preserve voltage stability. We will present basic concepts and provide references for detailed descriptions.

Hydro plant response. Large changes can be made quite fast at hydro generation. Response rates of 10–20% of capacity per minute are often quoted. Reference 25 describes a phase-plane controller installed at Hungry Horse power plant in Montana. Figure 5-12 shows a 75-MVA Hungry Horse generator going from no load to full load in about one minute.

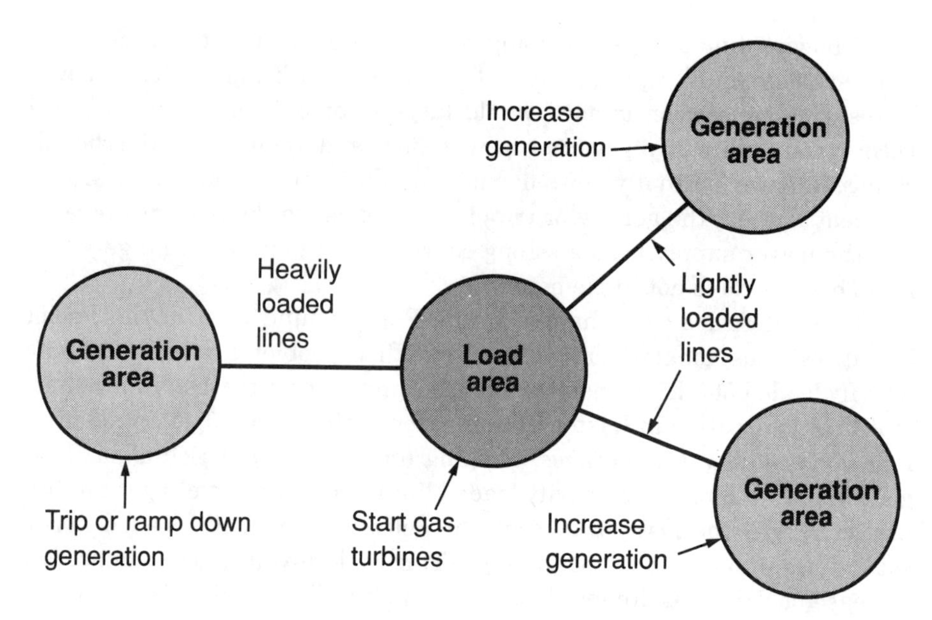

Fig. 5-11. Concepts of improving voltage stability by generation changes.

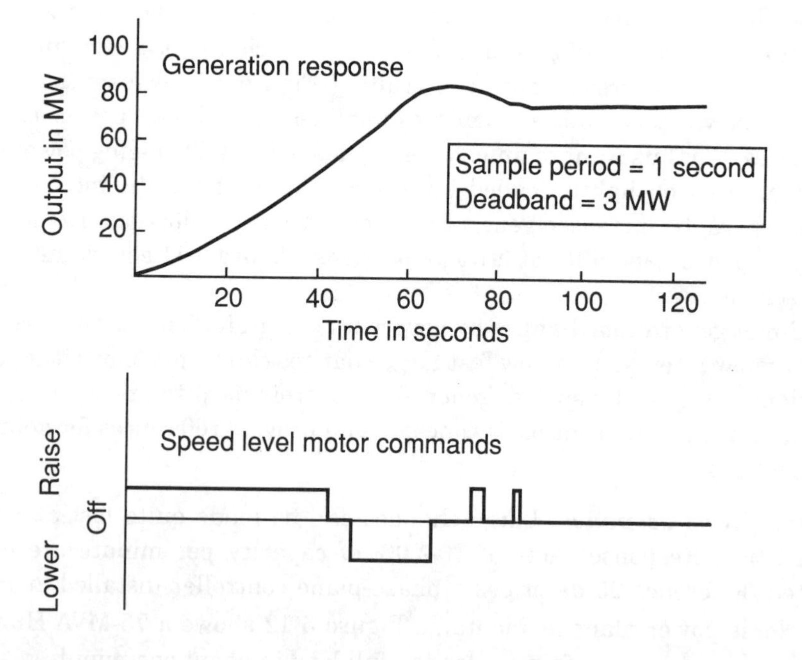

Fig. 5-12. Simulation (analog computer and actual digital controller) of Hungry Horse response at 100 MW/minute loading rate. © 1973 IEEE [25].

Thermal plant response. Response of spinning reserve at thermal plants depends on several factors: type of fuel, type of boiler (drum versus once-through), number of reheat stages, type of control (boiler following versus turbine following versus coordinated control). Response rates for 10–20% power changes of 2–3% of capacity per minute are often quoted. Larger changes require additional auxiliary units (coal mills, etc.). The rate limits are not only due to process lags but also due to permissible limits on temperatures, pressures, levels and corresponding stresses [26]. With careful equipment tuning and testing, a maximum rate of increase of 10% per minute might be achieved. Substantial improvement in voltage stability could be achieved with spinning reserve carried on several fast responding units.

Typical responses have been shown in several papers and reports [26–30]. Figure 5-13 shows typical response from an EPRI report. Recently there has been concern on deterioration of prime mover control [31]. One

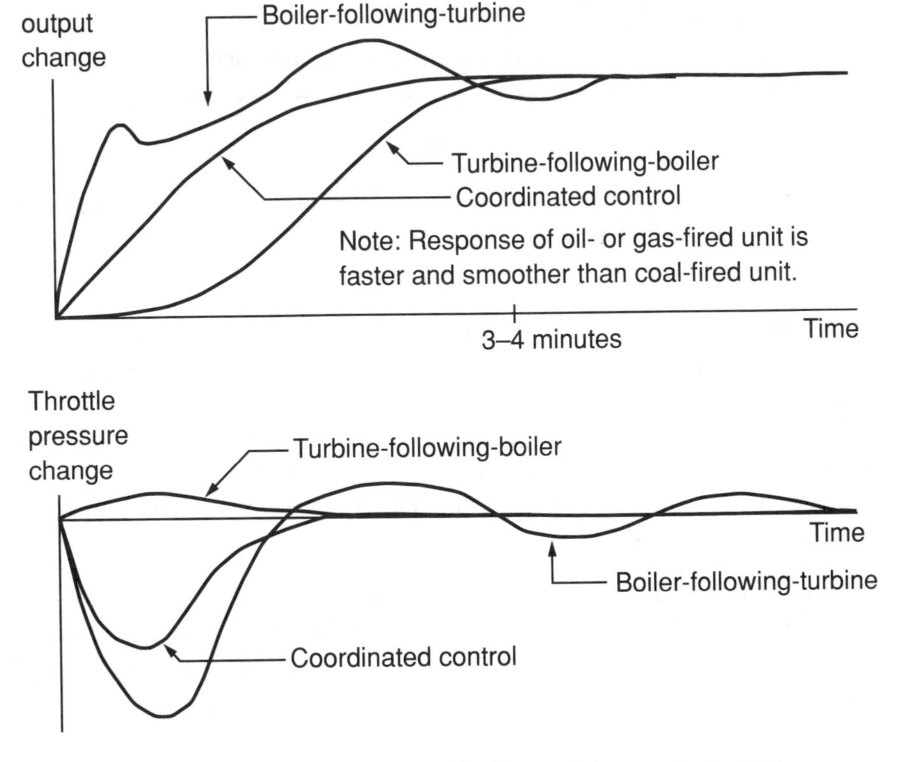

Fig. 5-13. Response for three types of boiler-turbine controls [30].

reason cited is increased use of sliding pressure control, which is similar to the turbine-following control shown on Figure 5-13.

There are special methods which may be implemented to achieve rapid power plant increase [32,33]. These include substantial throttling reserve, and temporary closure of feedwater heater extraction valves. Termuehlen and Gartner [32] described a combination of these methods to achieve a 20% load increase in less than one minute (Figure 5-14).

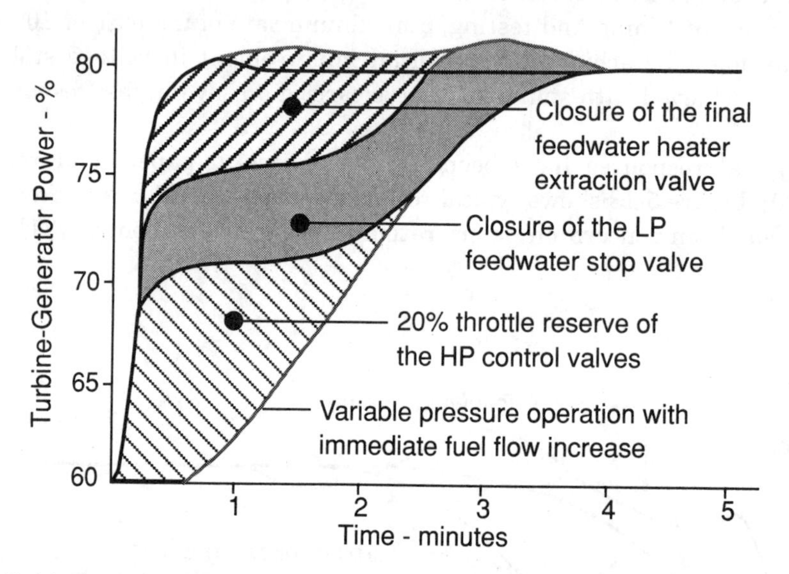

Fig. 5-14. Sustained fast unit load increase from 60% to 80%, utilizing various methods. © IEEE, 1980 [32].

Gas turbine response. Gas turbines can be brought on line and loaded relatively fast during an emergency. Time from cold start to full load is 5–20 minutes.

On May 20, 1986, a major voltage collapse in England was avoided by bringing 1000 megawatts of gas turbines on line following loss of six 400-kV circuits during a thunderstorm. The gas turbines were brought on line in five minutes [34].

Modeling of power plant response. Longer-term dynamic programs such as the EPRI ETMSP incorporate low-order models of energy supply systems (i.e., boilers, nuclear reactors, and combustion turbines). Models are described in references 29 and 30.

5.5 Automatic Generation Control (AGC)

We conclude this chapter with a brief description of automatic generation control in interconnected power systems. More detailed descriptions are available elsewhere, cf., references 35 and 36.

AGC is sometimes taken to include both *load-frequency control* and *economic dispatch*. We discuss only the faster load-frequency control which acts to maintain scheduled system frequency and scheduled net power interchange. Following a large disturbance (generation loss), AGC acts during a slowly developing voltage instability. As discussed above, the time frame is roughly 0.5 to 15 minutes following the disturbance. The power redistribution is based on an entirely different criterion than the faster governor redistribution.

AGC is a decentralized control operating in each control area. A control area generally corresponds to the service area of an individual utility. By decentralized, we mean that no signals are received from other control areas; the control is based on tie line power flow measurements and local system frequency measurement. Control is usually digital with 1–4 seconds compute cycle time.

Area frequency response characteristic. The combined response of governors and loads in an area to a power impact is termed the area frequency response characteristic. Governor responses are nominally based on steady state droop settings. Load responses to frequency excursions are due mainly to motor loads as indicated by the Pf parameter in Equation 4-1 and Table 4-1 (we often term the load response *damping*). There is also turbine damping because of the effect of abnormal speed on the steam or hydraulic torques operating on turbine blades.

The area frequency response characteristic, $\Delta P/\Delta\omega$, has the symbol β and is given by:

$$\beta = \frac{1}{R} + D \tag{5.1}$$

where R is composite governor droop of all generators in the area and D is composite load damping. If all units had active governors with 0.05 per unit droop, the $1/R$ term would be 20 per unit. The D term is normally much lower, 1–2 per unit. Thus governor response dominates the frequency recovery following generation loss. Since a drop in frequency (negative frequency deviation) results in generation increase, β is a negative number.

The area frequency response characteristic is routinely calculated at control centers following loss of generation in other control areas. The equation is:

$$\beta_{\text{meas}} = \frac{\Delta P_{\text{net int}}}{\Delta \omega} \tag{5.2}$$

where $\Delta P_{\text{net int}}$ is the measured net change or deviation in tie line power flow and $\Delta\omega$ is the measured change in system frequency (speed). The tie line power and the frequency measurements are made after primary prime mover control (governor) action is largely complete, and the system is temporarily quiescent. On Figure 5-9d, this is around fifteen seconds after the power impact.

From these routine calculations performed many times per year, the area responses have been found to be much less than expected from nominal governor droops. The overall eastern North American synchronous interconnection frequency response corresponds to about a 16% droop equivalent governor. The overall western North American interconnection frequency response corresponds to a 9–10% droop equivalent governor. The poor performances are due to block-loaded generators, governor deadbands and other nonlinearities, and very slow (turbine-following) prime mover control. The better performance of the western interconnection may be due to a larger proportion of hydro units with active governors.

Related to Equation 5.2, the system frequency or speed after prime mover control action is due to the frequency response characteristics of all control areas in the interconnection. We can write:

$$\Delta \omega = \frac{\Delta P}{\sum_i \beta_i} \tag{5.3}$$

where ΔP is the generation loss.

Example 5-4. The western North American interconnection capacity is about 100,000 MW. The generation loss is 1000 MW, or 1% of the interconnection capacity. Assuming the interconnection frequency response characteristic is -10 per unit (corresponding to 0.1 per unit or 10% droop equivalent governor), calculate the expected system frequency deviation after governor action.

Solution: Using equation 5.3, we calculate:

$$\Delta f = (60)\frac{0.01}{(-10)} = -0.06\ \text{Hz}.$$

Figure 5-15 shows actual western system response following 1000 MW generation loss on June 16, 1988. The frequency recovery is close to calculated value of -0.06 Hz.

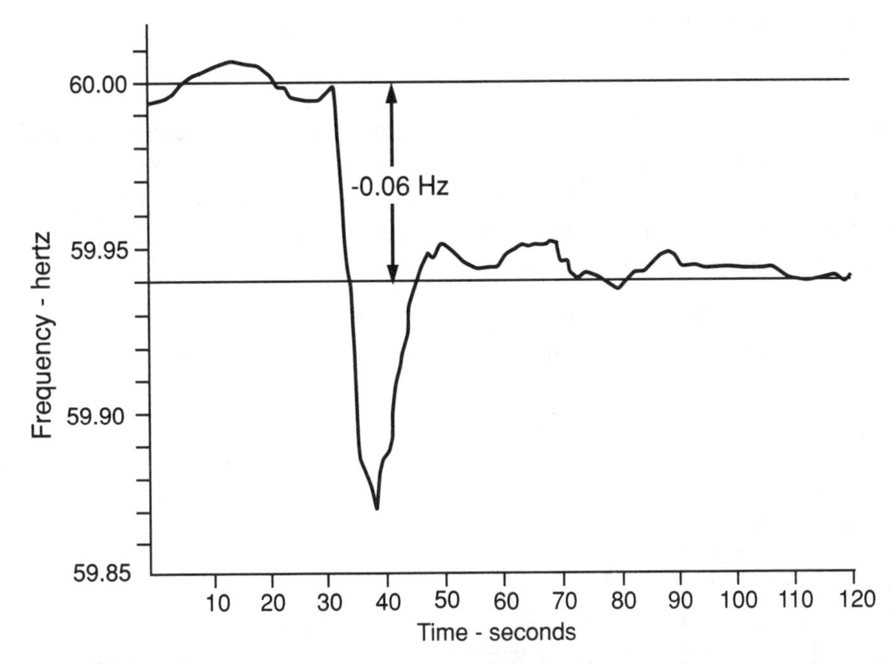

Fig. 5-15. System frequency response of loss of Jim Bridger units 1 and 2 (1000 MW) in the state of Wyoming. Recorded at Bonneville Power Administration control center in Vancouver, Washington at 1040 hours on 16 June 1988.

Area Control Error. The basic AGC algorithm calculates an *Area Control Error* based on pseudo steady-state concepts, namely that AGC is much slower than governor action. The equation is:

$$ACE = \Delta P_{\text{net int}} - B\Delta f \tag{5.4}$$

where B is *frequency bias* and is a negative number. The ACE sign convention is opposite the convention in control engineering—a positive ACE means lower generation. Internal loss of generation will cause both right hand side terms to be negative, resulting in negative ACE and commands to raise generation. The ACE is divided up among generating units according to participation factors.

From quasi steady-state analysis, perfect control for a power impact (step change in generation or load) results if $B = \beta$ in all control areas. By perfect control, we mean that $ACE = 0$ in all control areas except the area with the power impact; the ACE in the disturbed area is initially equal to the power impact. In external areas, the $B\Delta f$ term prevents premature undoing of governor assistance to the area suffering the generation loss.

Governor assistance gradually diminishes as frequency is returned to schedule.

Usually frequency bias is set only once per year based on calculated (measured) area frequency response characteristic during peak load conditions. For most disturbances, this results in some overcorrection and unnecessary control action. After some power exchanges, the power changes in external areas will be zero and the area with the generation loss will have completely covered the loss.

Example 5-5. For a two area system, show that perfect control action results with $B = \beta$. Show that unnecessary control action occurs with $B = 2\beta$.

The sketch below shows the model.

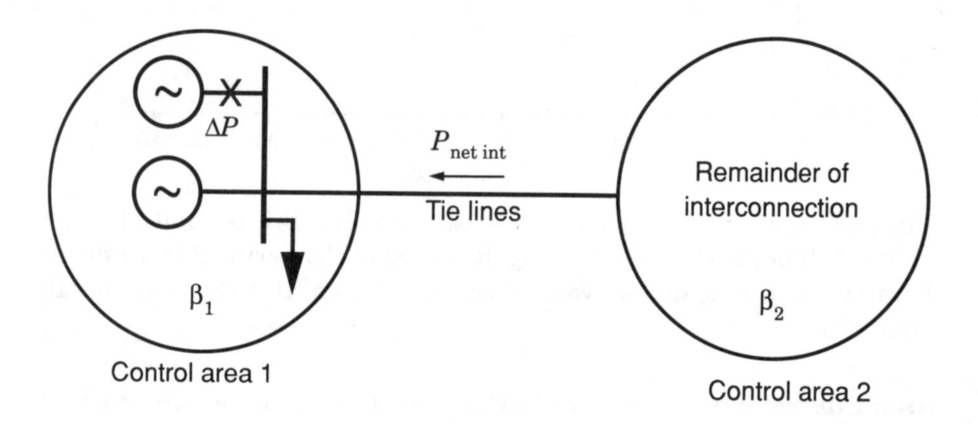

After governor action and with ΔP a negative value, the tie line power will be:

$$\Delta P_{\text{net int}} = \Delta P_{\text{area 2}} = \frac{\beta_2}{\beta_1 + \beta_2}\Delta P.$$

Using Equation 5.3, the frequency will be:

$$\Delta f = \frac{\Delta P}{\beta_1 + \beta_2}.$$

For $B = \beta$:

$$ACE_1 = -\frac{\beta_2}{\beta_1 + \beta_2}\Delta P - \beta_1\frac{\Delta P}{\beta_1 + \beta_2} = -\Delta P \text{ and}$$

$$ACE_2 = \frac{\beta_2}{\beta_1 + \beta_2}\Delta P - \beta_2 \frac{\Delta P'}{\beta_1 + \beta_2} = 0$$

Similarly for $B = 2\beta$:

$$ACE_1 = -\frac{\beta_2}{\beta_1 + \beta_2}\Delta P - 2\beta_1 \frac{\Delta P}{\beta_1 + \beta_2} = -\Delta P\left(\frac{2\beta_1 + \beta_2}{\beta_1 + \beta_2}\right)$$

$$ACE_2 = \frac{\beta_2}{\beta_1 + \beta_2}\Delta P - 2\beta_2 \frac{\Delta P}{\beta_1 + \beta_2} = -\Delta P\left(\frac{\beta_2}{\beta_1 + \beta_2}\right)$$

Voltage instability usually occurs during heavy load conditions. We see that only small temporary generation increases will be made in external areas. This is completely different from "governor" power flow which is due to frequency response characteristics of areas.

Emergency actions, AGC suspension or rescheduling. For large disturbances, the area suffering the generation loss will not be able to cover the loss and power interchange schedule changes will be made. For voltage stability, the additional power import should be scheduled over lightly loaded lines.

AGC may be suspended for major disturbances. For example, a large frequency change indicates electrical islanding; AGC should not try to control interchange over tie lines that may be open. AGC may be suspended based on direct detection of outage of a major tie line. Finally, AGC may be suspended for intentional generation tripping; for voltage stability, this prevents AGC from undoing the generation tripping.

An alternative to AGC suspension for detection of tie lie outages or for intentional generator tripping is rescheduling of power interchanges, keeping AGC in service.

As described above, special fast generation control may aid voltage stability. Automatic Generation Control facilities may be used for this emergency control.

References

1. B. M. Weedy, *Electric Power Systems*, Third Edition, John Wiley & Sons, 1979.
2. A. E. Fitzgerald, C. Kingsley, Jr., and S. D. Ulmans, *Electric Machinery*, Fifth Edition, McGraw-Hill, 1990.
3. P. Kundur, *Power System Stability and Control*, McGraw-Hill, 1993.
4. H. M. Rustebakke, editor, *Electric Utility Systems and Practices*, Fourth Edition, John Wiley & Sons, 1983.
5. A. Capasso and E. Mariani, "Influence of Generator Capability Curves Representation on System Voltage and Reactive Power Control Studies," *IEEE Trans-*

actions on Power Apparatus and Systems, Vol. PAS-97, No. 4, pp. 1036–1041, July/August 1978.

6. W. R. Lachs, "System Reactive Power Limitations," IEEE/PES paper A 79 015-9, IEEE/PES Winter Meeting, 1979.
7. CIGRÉ Task Force 38-02-10, *Modelling of Voltage Collapse Including Dynamic Phenomena*, 1993.
8. A. Calvaer, "Voltage Stability and Collapses: a Simple Theory Based on Real and Reactive Currents," *Revue Generale de l'electricite*, No. 7, pp. 1–17, September 1986.
9. P. B. Johnson, S. L. Ridenbaugh, R. D. Bednarz, and K. G. Henry, "Maximizing the Reactive Capability of AEP Generating Units," *Proceedings of the American Power Conference*, pp. 373–377, April 1990.
10. A. S. Rubenstein and W. W. Walkley, "Control of Reactive KVA with Modern Amplidyne Voltage Regulators," *AIEE Transactions*, pp. 961–970, December 1957.
11. A. S. Dehdashti, J. F. Luini, and Z. Peng, "Dynamic Voltage Control by Remote Voltage Regulation for Pumped Storage Plants," *IEEE Transactions on Power Systems*, Vol. 3, No. 3, pp. 1188–1192, August 1988.
12. S. Koishikawa, S. Ohsaka, M. Suzuki, T. Michigami, and M. Akimoto, "Adaptive Control of Reactive Power Supply Enhancing Voltage Stability of a Bulk Power Transmission System and a New Scheme of Monitor on Voltage Security," *CIGRÉ*, paper 38/39-01, 1990.
13. D. Brandt, R. Wachal, R. Valiquette, and R. Wierckx, "Closed Loop Testing of a Joint VAR Controller Using a Digital Real-Time Simulator," *IEEE Transactions on Power Systems*, Vol. 6, No. 3, pp. 1140–1146, August 1991.
14. R. Wachal, D. P. Bradt, and R. Valiquette, "Dorsey HVDC Station Joint Var Controller, A Programmable Controller Project," *Proceedings of the Third HVDC System Operation Conference*, Winnipeg, pp. 133–138, 2–15 May 1992.
15. J. P. Paul, C. Corroyer, P. Jeannel, J. M. Tesseron, F. Maury, and A. Torra, "Improvements in the Organization of Secondary Voltage Control in France," *CIGRÉ*, paper 38/39-03, 1990.
16. V. Archidiacono, S. Corsi, A. Natale, C. Raffaelli, and V. Menditto, "New Developments in the Application of ENEL Transmission System Voltage and Reactive Power Automatic Control," *CIGRÉ*, paper 38/39-06, 1990.
17. J. P. Piret, J. P. Antoine, M. Stubbe, N. Janssens, and J. M. Delince, "The Study of a Centralised Voltage Control Method Applicable to the Belgian System," CIGRÉ, paper 39-201, 1992.
18. M. S. Baldwin and D. P. McFadden, "Power System Performance as Affected by Turbine-Generator Controls During Frequency Disturbances," *IEEE Transactions on Power Apparatus and Systems*, Vol. PAS-100, No. 5, May 1981.
19. IEEE Committee Report, *IEEE Guide for AC Generator Protection*, ANSI/IEEE C37.102-1987.
20. IEEE Std. 122-1985, *IEEE Recommended Practice for Functional and Performance Characteristics of Control Systems for Steam Turbine-Generator Units*, IEEE, 1985.
21. M. Lotfalian, R. Schlueter, D. Idizior, P. Rusche, S. Tedeschi, L. Shu, and A. Yazdankhah, "Inertia, Governor, and AGC/Economic Dispatch Load Flow Simulations of Loss of Generation Contingencies," *IEEE Transactions on Power Apparatus and Systems*, Vol. PAS-104, No. 11, pp. 3020–3028, November 1985.

22. P. M. Anderson and A. A. Fouad, *Power System Control and Stability*, The Iowa State University Press, Ames, Iowa, 1977.
23. R. Rudenberg, *Transient Performance of Electric Power Systems: Phenomena in Lumped Networks*, McGraw-Hill, New York, 1950. (MIT Press, Cambridge, Mass., 1967.)
24. IEEE Committee Report, *Voltage Stability of Power Systems: Concepts, Analytical Tools, and Industry Experience*, IEEE/PES 90TH0358-2-PWR, 1990.
25. D. N. Scott, R. L. Cresap, R. F. Priebe, D. E. Tehrink, and K. A. Takeuchi, "Closed Loop Digital Automatic Generation Controller," IEEE/PES paper C 73 518-8, 1973.
26. IEEE Committee Report, "MW Response of Fossil Fueled Steam Units," *IEEE Transactions on Power Apparatus and Systems*, Vol. PAS-92, No. 2, March/April 1973. Reprinted in reference 28.
27. C. Concordia, F. P. de Mello, L. K. Kirchmayer, and R. P. Schulz, "Effect of Prime-Mover Response and Governing Characteristics on System Dynamic Performance," *Proceedings of the American Power Conference*, 1966. Reprinted in reference 25.
28. IEEE Committee Report, *Symposium on Power Plant Response*, IEEE publication TH0105-7-PWR, 1983.
29. IEEE Committee Report, "Dynamic Models for Fossil Fueled Steam Units in Power System Studies," *IEEE Transactions on Power Systems*, Vol. 6, No. 2, pp. 753–761, May 1991.
30. Ontario Hydro, *Long-Term Dynamic Simulation: Nuclear and Thermal Power Plant Models*, EPRI Final Report TR-101765, December 1992.
31. EPIC Engineering, Inc., *Impacts of Governor Response Changes on the Security of North American Interconnections*, EPRI Final Report TR-101080, October 1992.
32. H. Termuehlen and G. Gartner, "Sustained Fast Turbine-Generator Load Response in Fossil-Fueled Power Plants," *IEEE Transactions on Power Apparatus and Systems*, Vol. PAS-100, No. 5, pp. 2495–2503, May 1981.
33. Power Technologies Incorporated, *Technical Limits to Transmission System Operations*, EPRI Final Report EL-5859.
34. M. G. Dwek, Study Group 38 discussion, *Proceedings of 33rd CIGRÉ Session*, Vol. II, 1988.
35. A. J. Wood and B. L. Wollenberg, *Power Generation Operation and Control, John Wiley & Sons*, New York, 1984.
36. F. P. de Mello and J. M. Undrill, "Automatic Generation Control," *IEEE Tutorial Course 77 TUDO 010-9-PWR*. Reprinted in reference 28.

6

Simulation of Equivalent Systems

Seek simplicity—and distrust it.
Alfred North Whitehead.

We have described characteristics and modeling of power system equipment. Now we use the models for simulation. This chapter describes steady state and dynamic simulation of small equivalent systems. These systems provide considerable insight into voltage stability phenomena. Results are easily tractable.

The first system has an infinite source and only load and reactive power compensation effects are studied. Both load and generation effects are studied in the second example.

Before starting this chapter, you may wish to look at Appendix C, *Power Flow Simulation Methodology.*

6.1 Equivalent System 1: Steady-State Simulation

Figure 6-1 shows the equivalent system. A 600 MW load is served from a large system over two 230-kV lines, each 113 km long. Generator current limiting is not considered because of the infinite source system assumption. The system is heavily stressed and heavily shunt compensated—possibly because of difficulty in building new transmission lines.

The load is half motor and half resistive. The voltage at the load is controlled by an LTC transformer. The resistive load is constant energy (thermostatically controlled). The motor load is 80% shunt capacitor compensated. The receiving-end 230-kV bus has additional shunt compensation.

Steady-state simulation. Using quasi-dynamic analysis, we compute *V–Q* curves at different points in time. The points in time are before and

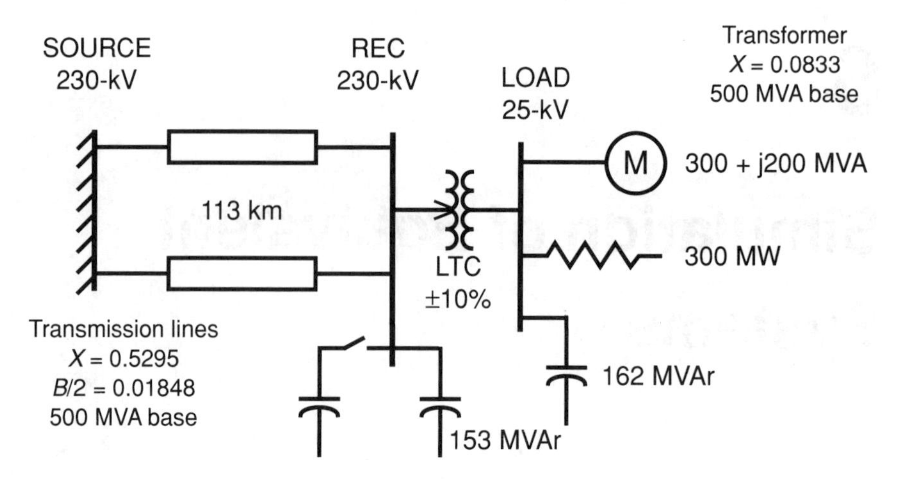

Fig. 6-1. Equivalent system 1. Two 113 km, 230-kV lines deliver 600 MW to load that is half induction motor (80% shunt compensated) and half resistive. Prior to outage of a line, SOURCE and REC busses are at 1.05 per unit voltage, and the LOAD bus is at 1.0 per unit voltage.

after outage of one of the 230-kV lines. The outage severely stresses the system and both stability and thermal limits are approached.

Because of the motor load, the analysis is highly approximate—but instructive. Section 6.2 describes more exact dynamic analysis.

Figure 6-2 shows the *V–Q* plots computed using a power flow program. Five system characteristics are shown. The receiving-end 230-kV bus is the test bus for the fictitious synchronous condenser. Three shunt capacitor bank characteristics are shown. The system characteristics are:

1. Characteristic (1)—pre-disturbance system. Point A, the intersection of the system characteristic and the 153 MVAr capacitor bank characteristic, is the operating point. The test bus voltage is 1.05 per unit. Voltage at the load is held to 1.0 per unit by the LTC transformer.
2. Characteristic (2)—immediately following the outage. The induction motor load initially has an impedance characteristic. Point B is the operating point.
3. Characteristic (3)—prior to LTC control. Constant real power is now assumed for the motor.
4. Characteristic (4)—after LTC control. For voltages below about 0.9 per unit, the LTC is at maximum boost limit and the resistive load is voltage dependent.

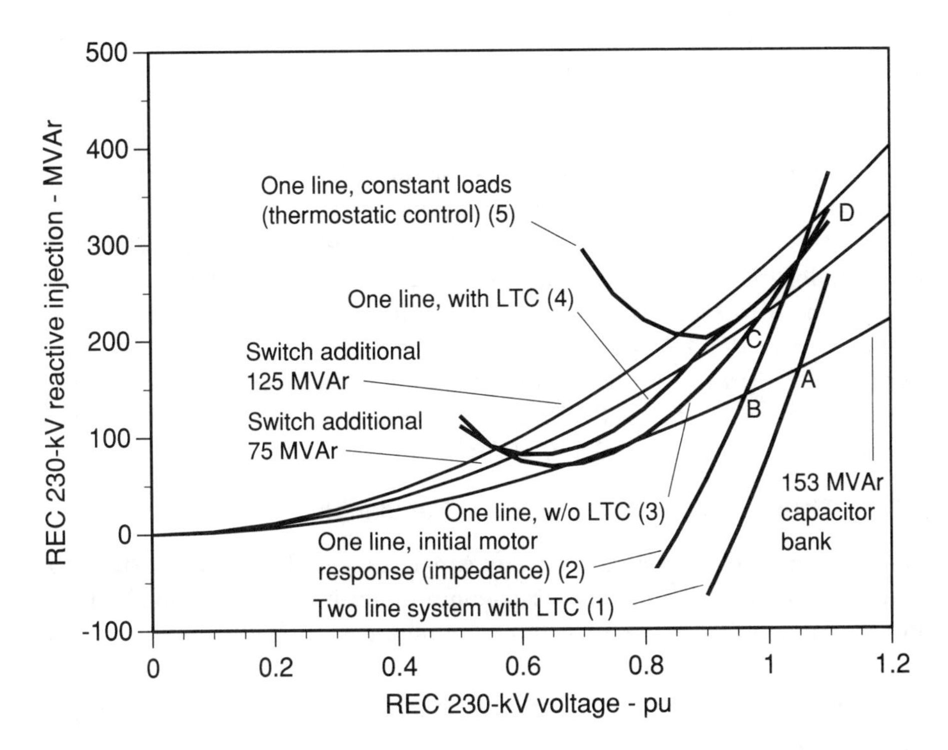

Fig. 6-2. *V–Q* curves for outage of one line and for several time frames.

5. Characteristic (5)—load is constant because of LTC and thermostatic control.

Transient voltage stability is a problem. We barely have an operating point (intersection of system characteristic (3) and the 153 MVAr capacitor bank characteristic). The operating point near 0.7 per unit voltage is unacceptable and the motor would stall. If the disturbance involved a short circuit, collapse would be very fast as the motor would not reaccelerate.

Switching a shunt capacitor bank on could at least temporarily preserve stability. The switching needs to be fast because the transition from system characteristic (2) to characteristic (3) takes only a few seconds. Switching a 75 MVAr capacitor results in a temporary operating point at Point C near one per unit voltage. Voltage at the load is below normal. Collapse would occur in a minute or two since system characteristic (4) does not intersect the capacitor characteristic. Switching a 125 MVAr capacitor bank on stabilizes the system near Point D, but voltage would be high.

Dynamic simulation is needed to determine capacitor bank size and the switching speed. Considerations include short circuit disturbances, performance for various operating conditions, high voltage problems, and whether or not the source system is completely stiff.

Figure 6-2 shows that an intersection of system and capacitor bank characteristics between 1.0 and 1.05 per unit voltage is not well defined. A better solution is necessary. Solutions include a load-area generator, a synchronous condenser, or a static var compensator.

For the same system characteristics, Figure 6-3 shows application of an SVC. Following the outages, the operating point would immediately move from Point A to Point B. Within a few cycles, however, the system would stabilize near Point C. The response of the SVC is faster than the motor dynamics. Very little tap changing results and the operating point is well defined. Retaining the 153 MVAr fixed capacitor bank, the SVC control range should be about +150, -50 MVAr.

6.2 Equivalent System 1: Dynamic Simulation

We now simulate the line outage and capacitor switching using a transient stability program. Transient voltage/motor stability will be studied, including the effect of short circuits on motor reacceleration. The motor equiva-

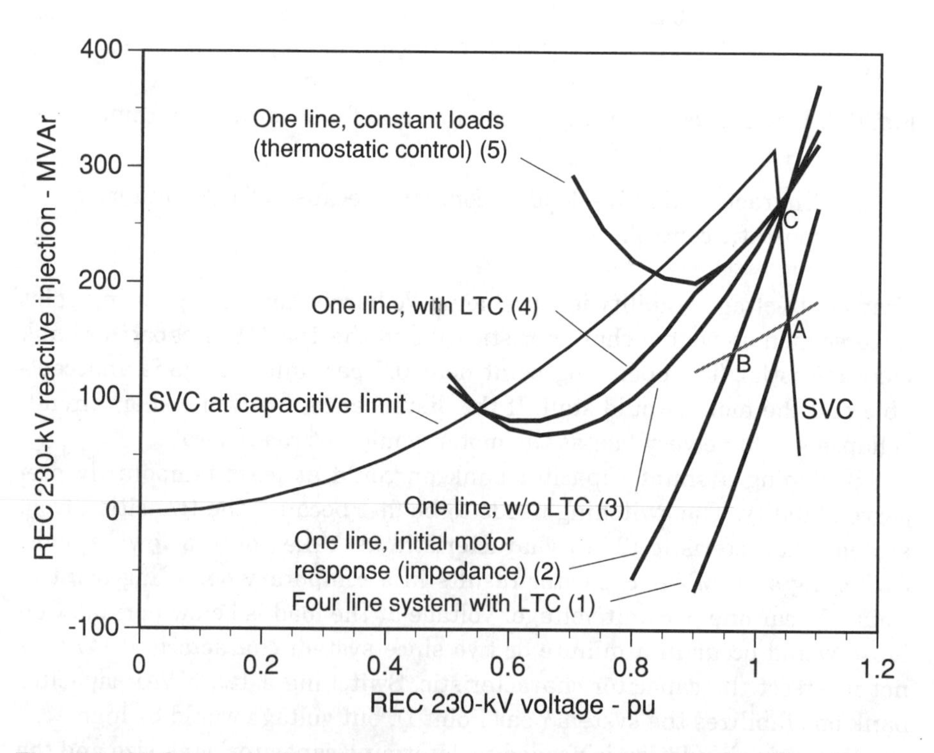

Fig. 6-3. *V–Q* curves for outage of one line and for several time frames. Application of static var compensator.

lent corresponds to the small industrial induction motor listed in Table 4-1. The equivalent circuit and other parameters are repeated in Table 6-1.

Table 6-1

R_s	X_s	X_m	R_r	X_r	H	A	LF
0.031	0.1	3.2	0.018	0.18	0.7	1	0.6

The equivalent circuit parameters are in per unit on the base of the equivalent motor which is 500 MVA. LF is the load factor of 300 MW/500 MVA. $A = 1$ means the load torque is proportional to the square of the speed.

The equivalent motor represents many motors that respond similarly to transmission system disturbances. Maintaining stability with a high proportion of motor load and a weak transmission system is difficult. The stability limit for even non-fault disturbances may be lower than the static power transfer limit.

Torque-speed curves. Before describing time domain simulations we can assess stability by torque-speed curves. Torque-speed curves are usually given for constant motor terminal voltage. This is appropriate for studying a single motor fed from a strong power system. In our case, however, the aggregated or equivalent motor sees a fixed voltage only at the infinite bus. In computing the torque-speed curves, the thévenin equivalent circuits shown on Figure 6-4 apply. The thévenin circuits are based on the Figure 6-1 system and the motor equivalent circuit. The 300 MW resistive load equivalent is part of the thévenin circuits. Torque is calculated for different values of slip.

Furthermore, results with equivalent motor modeling should be interpreted cautiously. A single stall-prone motor (heavily loaded, low inertia, constant torque load) at the end of a long feeder could stall and cause cascade stalling of other motors.

Figure 6-5 shows the torque-speed curves. Curves for three conditions are shown:

1. Fixed motor terminal voltage (for reference only);
2. Two line pre-disturbance system; and
3. Outage of one line, additional 125 MVAr capacitor bank switched in.

We see that motor stability margin is very low even in the pre-disturbance system. This is the direct consequence of the point of fixed voltage being at the far end of very heavily loaded transmission lines. A small increase in

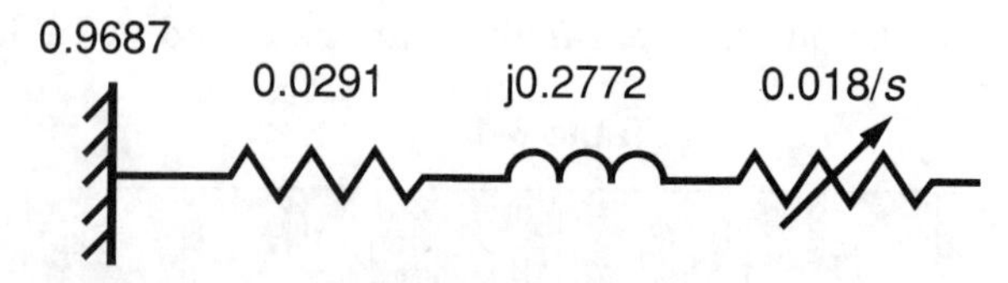

(a) Thévenin equivalent with motor terminal voltage fixed at one per unit.

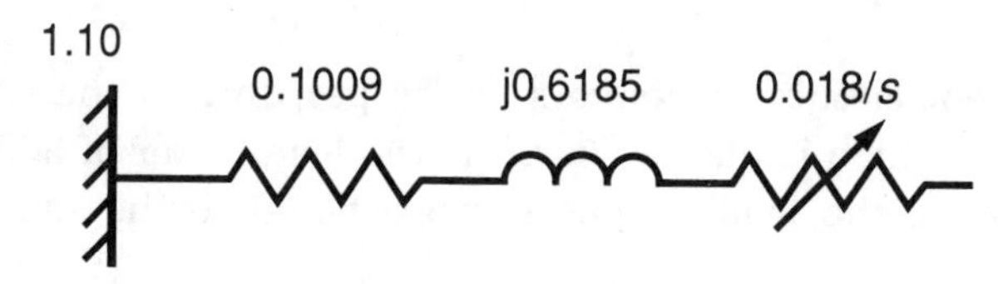

(a) Thévenin equivalent for two line system.

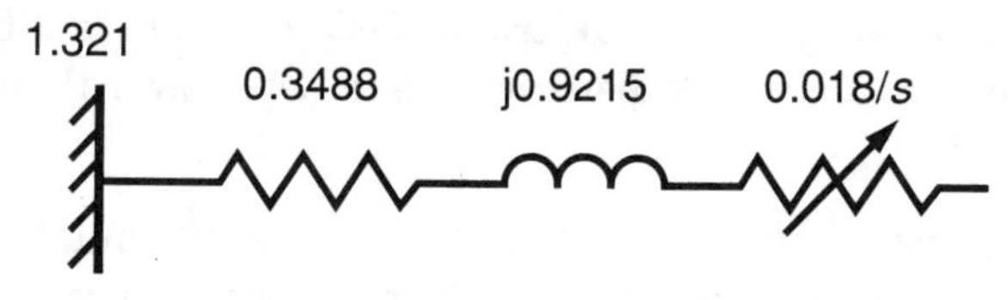

(a) Thévenin equivalent for outage of one line and with additional 125 MVAr capacitor bank.

Fig. 6-4. Thévenin equivalents looking into system from motor rotor. Source system shown on Figure 6-1. Base power is the 500-MVA motor rating.

the motor load, or increase in motor load factor would cause instability. We can also imagine a short circuit deccelerating the motor to a speed where it does not reaccelerate following fault clearing. Note the small inertia constant of 0.7 MW-seconds/MVA. Regulation of load end voltage by an SVC or synchronous condenser would, of course, improve the torque-speed characteristic.

Dynamic simulation for three types of reactive compensation. The motor representation is the third order dynamic model described in Chapter 4. We apply a three-phase fault at the midpoint of one line. The fault is cleared in four cycles by permanently opening the line. Stability is marginal. We simulate three cases with different reactive power compensation applied at the REC 230-kV bus:

1. 125 MVAr mechanically switched capacitor inserted 0.1 seconds after fault clearing.

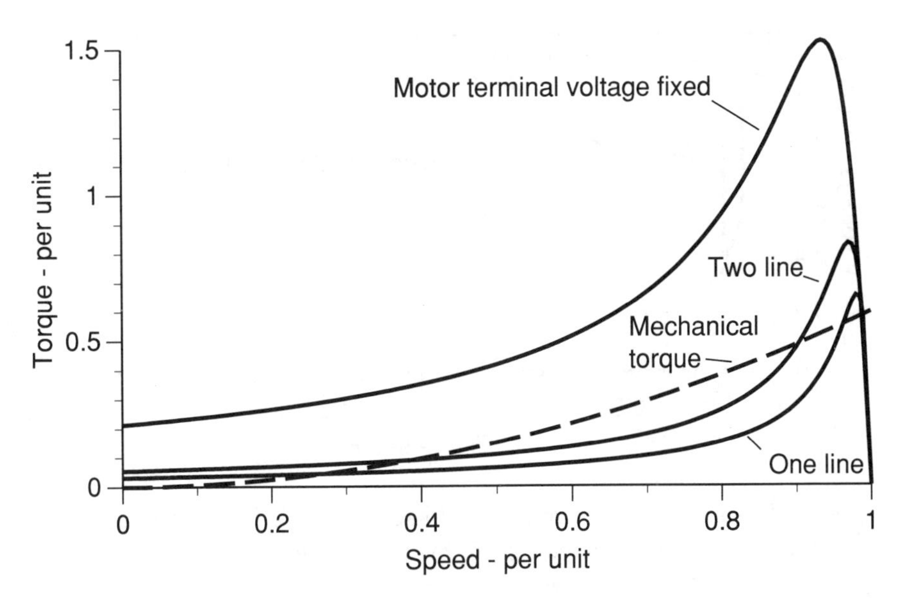

Fig 6-5. Torque-speed curves of equivalent induction motor for the théve-nin equivalent circuits of Figure 6-4.

2. +125 MVAr, -75 MVAr static var compensator. A 50 millisecond first order lag represents the principal dynamics. A 2% slope is used corresponding to dynamic gain of 50 per unit.
3. 125 MVAr synchronous condenser. The machine parameters are taken from reference 2. A powerful static excitation system is used. Line drop compensation is used for fast regulation of the REC 230-kV bus. (Both the SVC and synchronous condenser regulate the 230-kV voltage.)

Figure 6-6 shows voltage at the REC 230-kV bus for the three cases. All cases are stable. We make the following remarks.

- The MSC is the least expensive option but has important performance disadvantages. Stability margin is smaller than the other options and high voltage results in the post-disturbance steady state (as predicted by the *V–Q* curves). Transmission line reclosing would cause high temporary overvoltages until the capacitor bank is switched off. The switching time is critical—there is no time to wait for a reclosing attempt.
- The SVC is blocked during the fault on period, but immediately goes to full capacitive boost after fault clearing (in some SVCs there is delay of perhaps 30 milliseconds in deblocking after

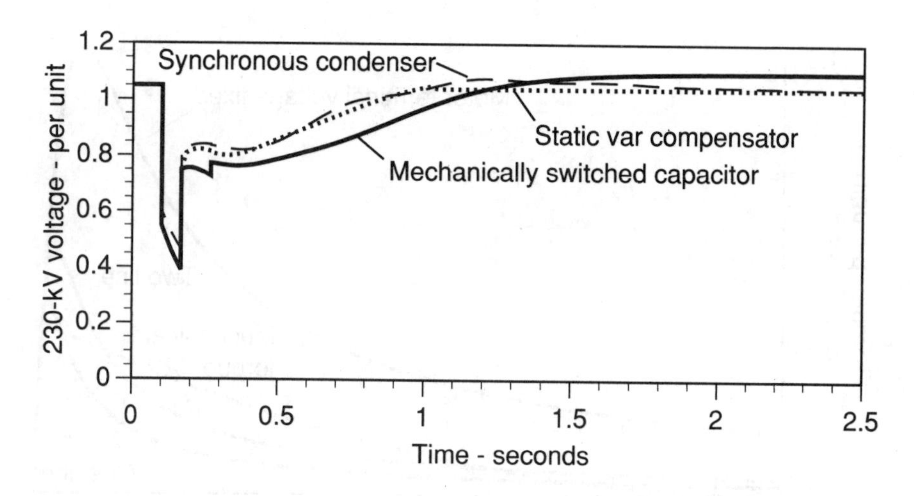

Fig. 6-6. Voltage recovery for three-phase midpoint fault with permanent line outage. Three reactive power compensation options.

fault clearing). Output remains at full boost until voltages recover. Response is fast and well damped. SVC dynamics are faster than the induction motor dynamics. Final output is about 100 MVAr, providing 25 MVAr of reserve in the boost direction.

- The synchronous condenser provides slightly greater stability margin than the SVC. The fault voltage dip is reduced and the substantial temporary overload capability is used. Because of synchronous machine time lags and inertial swings, the response is more oscillatory than the SVC option. The synchronous condenser response speed is similar to that of the induction motor. Similar dynamic interaction would occur if generators were represented. The SVC is more cost-effective because of lower life-cycle costs.

References 3, 4, and 5 report somewhat similar transient voltage stability studies of induction motors and static var compensators.

6.3 Equivalent System 2: Steady-State Simulation

Figure 6-7 shows a more complicated equivalent system [1]. Two loads are feed from a 500-kV bus in the load area. Industrial load is served directly via a LTC transformer. Residential and commercial load is served via two LTC transformers and an equivalent for subtransmission impedance. The load area is heavily shunt compensated. The load area also includes a 1600 MVA equivalent generator.

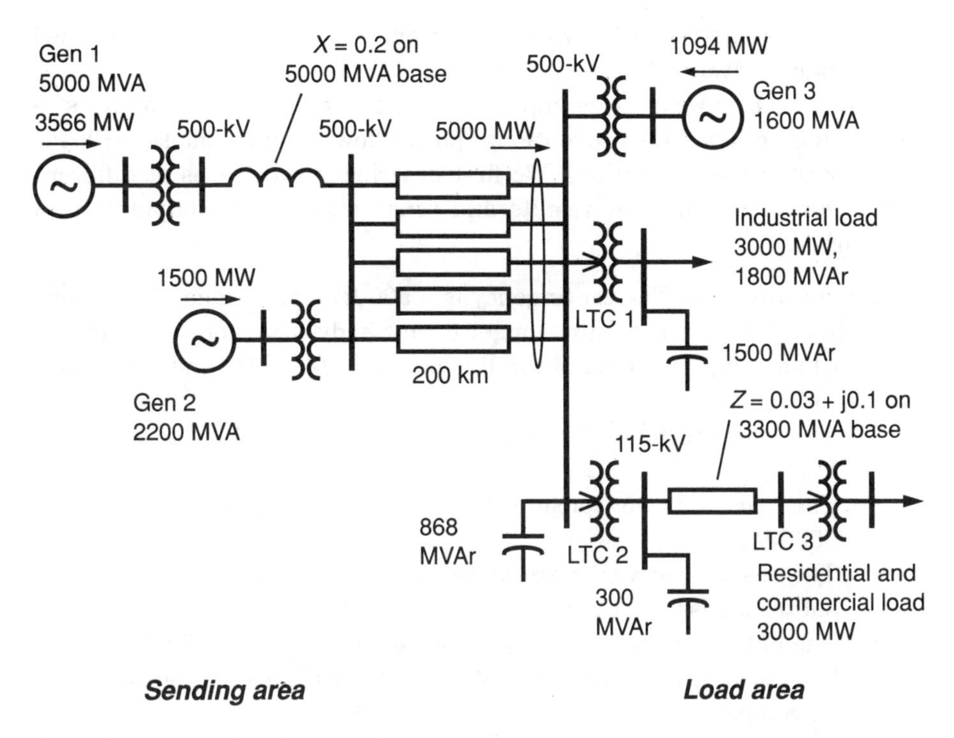

Fig. 6-7. Equivalent System 2.

Two remote generators deliver 5000 MW to the load area over five 500-kV lines.

Appendix E provides the steady-state and dynamic data for Equivalent System 2.

Base conditions for power flow simulations are:

- LTCs 1 and 3 are automatically controlled to regulate the low-side voltage. LTC 2 has fixed tap. Continuous rather than discrete taps are modeled for the LTC transformer equivalents.
- The industrial load active power is constant. The reactive power is constant current ($Q \propto V^1$).
- The residential and commercial load is unity power factor. Half of the load is constant and half is resistive For small voltage changes this is the same as a constant current characteristic.
- All generator voltage regulators control terminal voltage. Generators are initially operated near unity power factor.

The combined load is voltage insensitive—75% constant and 25% resistive. This is onerous for voltage stability. Since most of the load is motor, steady-state analysis must be used with caution and results should be confirmed by dynamic analysis.

In this section we analyze important aspects of longer-term voltage stability using an available conventional power flow program models (PV/PQ bus generator representation). We first describe outage cases at different load levels. We then show results and sensitivities from computation of *V–Q* curves.

Outage of one line. The contingency is a 500-kV line outage. We assume transient stability and examine longer term steady-state behavior. Let's do some bookkeeping: the effect of the line outage *before* LTC and area interchange control is:

- The load-end 500-kV bus voltage drops from 540 kV to 529.5 kV.
- The reactive power consumption of the 500-kV lines increase by 652 MVAr.
- The reactive power consumption of the transformers and the subtransmission equivalent increase by 20 MVAr
- The output of the shunt capacitor banks decrease by 75 MVAr. This includes the net effect of the industrial reactive load and its compensation.
- Reactive power output of Gen 1, Gen 2, and Gen 3 increase by 192 MVAr, 234 MVAr, and 319 MVAr respectively. The total increase of 745 MVAr equals the reactive power loss increase and shunt capacitor bank output reduction. It's easier to keep track of the generator reactive power.
- The load decreases by 62 MW, but the active power losses increase by 18 MW.

The effect of the line outage *after* LTC and area interchange control is:

- The load-end 500-kV bus voltage drops from 540-kV (base case) to 528.9 kV.
- The reactive power consumption of the 500-kV lines increase by 716 MVAr over the base case.
- The reactive power consumption of the transformers and the subtransmission equivalent increase by 47 MVAr
- The output of the shunt capacitor banks decrease by 54 MVAr.
- Reactive power output of Gen 1, Gen 2, and Gen 3 increase by 222 MVAr, 255 MVAr, and 338 MVAr respectively. The total increase of 815 MVAr equals the reactive power loss increase and shunt capacitor bank output reduction.

Load and transfer increases. To stress the five line system, we increase the residential/commercial load, dispatching generation from Gen 1. The load increase is half constant and half resistive at unity power factor. We then take an outage of one line. We test system robustness for the first contingency by applying a 100 MVAr constant load at the residential/commercial load bus. The 100 MVAr load allows us to compute sensitivities to small changes.

Figure 6-8 shows the resulting *P–V* curve for increase in load and transfer. The ordinate is the load area 500-kV bus voltage. This bus feeds both loads. Curves are shown for the five line system with LTC control, the system with outage before LTC control, and the system with outage after LTC control. For all three conditions, area interchange control is active.

Point D at 5400 MW transfer (line outage with LTC control) is an unacceptable operating point. The 500-kV voltage has dropped 20 kV (4%) because of the outage. More importantly, Gen 3 is at its reactive limit and Gen 2 is only 114 MVAr away from its limit. LTC 3 is at its limit. Applying the 100 MVAr load causes power flow divergence. At Point D, the angle between Gen 1 and Gen 3 is 29.4°.

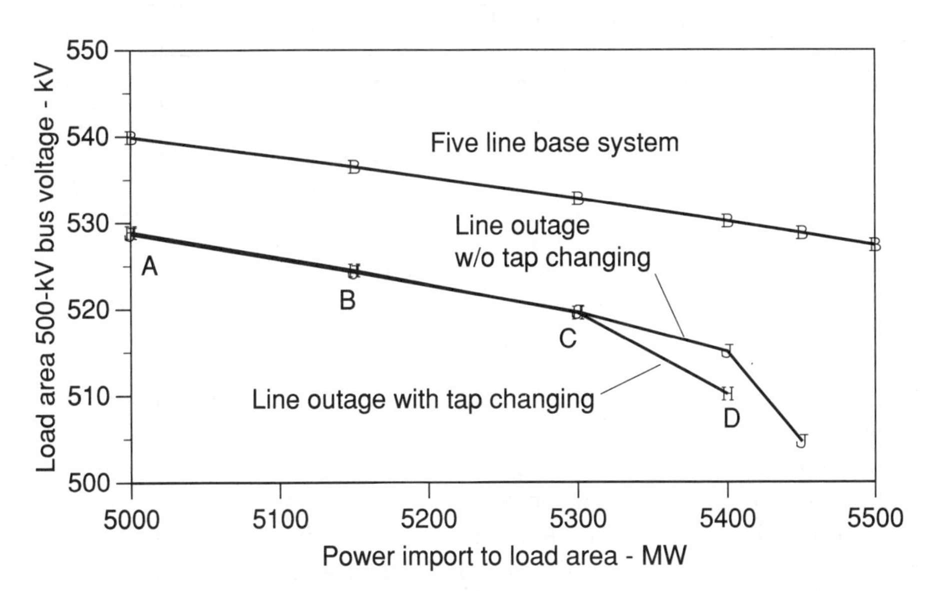

Fig. 6-8. Effect of increasing residential and commercial load with power import from sending area. For each base case load level, a one line outage contingency is simulated with and without tap changing. For outage without tap changing, power flow diverges for 5500 MW transfer. For outage with tap changing, power flow diverges for 5450 MW transfer.

Sensitivities. Sensitivities for small changes can be computed. Although some power flow programs compute some sensitivities from a solved case as an output option (Appendix B), we run additional cases so that any sensitivity may be determined. We apply a constant (voltage insensitive) 100 MVAr load at the residential/commercial load bus and note the changes from the previous case. Bus, branch, and generator sensitivities (participations) can be computed. Generator reactive power sensitivities are particularly interesting.

For the 100 MVAr load additions, Table 6-2 shows the sensitivities for Points A, B, C, and D of Figure 6-8. For Point C, a 10 MVAR load addition is used, since a 100 MVAR addition causes Gen 3 to hit its reactive limit and also causes LTC 3 to hit its tap limit. For Point D, a 10 MVAr load addition allows power flow convergence.

Table 6-2

	$\frac{\Delta V_{\text{Load 500-kV}}}{\Delta Q_{\text{Load}}}$	$\frac{\Delta Q_{\text{Gen 1}}}{\Delta Q_{\text{Load}}}$	$\frac{\Delta Q_{\text{Gen 2}}}{\Delta Q_{\text{Load}}}$	$\frac{\Delta Q_{\text{Gen 3}}}{\Delta Q_{\text{Load}}}$	$\frac{\Delta Q_{\text{Gen total}}}{\Delta Q_{\text{Load}}}$
Point A	-0.03[a]	0.332	0.386	0.911	1.629
Point B	-0.03	0.054	0.411	0.964	1.729
Point C	-0.03	0.38	0.43	1.01	1.82
Point D[b]	-0.14	1.56	1.81	0[c]	3.37

a. kV/MVAr
b. Gen 3 at reactive power limit, LTC 3 at boost limit.
c. Gen 3 at reactive power limit.

The last column is the total change in generator reactive power for the load change. This is the *Voltage Collapse Proximity Indicator*, *VCPI*, introduced in Chapter 2 (Example 2-3). For Point D, 3.37 MVAr of reactive generation is required for each additional MVAr of load at the residential/ commercial load bus. Once Gen 3 is at its reactive limit, the VCPI will increase rapidly—theoretically to infinity at the maximum power point. Since, at Point D, LTC 3 is at its boost limit (load is voltage sensitive), the VCPI increase from Point C is moderate. From Point A to Point D, the VCPI approximately doubles.

V–Q curves. *V–Q* curve computation is automated in power flow programs such as EPRI's VSTAB program. Sensitivities to model and control changes over a wide range of stress levels are rapidly computed. The load area 500-

kV bus is the test bus for stressing the transmission system. This is the only transmission system bus in the load area. Points are computed at five kV intervals between 475 and 545 kV. All cases are for 6000 MW of load and 5000 MW of imports. Area interchange control is active except as noted.

V–Q curves—reference case, outage of one line. One of the 500-kV lines is out-of-service. This reference case has LTC transformers 1 and 3 controlling voltage. LTC transformer 2 is operating with fixed tap. All three generators control terminal voltage. Figure 6-9 shows the *V–Q* curve and the 868 MVAr capacitor bank characteristic. Also shown are the reactive power outputs of the three generators.

Gen 3 reaches its reactive power limit of 700 MVAr at a scheduled voltage of 515 kV. Gen 2 reaches its reactive power limit of 725 MVAr at 490 kV. LTC 1 is at limit at 500 kV and LTC 3 is at limit at 495 kV. The change in *V–Q* curve slope when Gen 3 reaches its reactive power limit is especially notable. Because part of the load becomes voltage sensitive with LTC 3 on limit, the *V–Q* curve does not reach a minimum and turn upward.

The operating point is the intersection of the *V–Q* curve and the capacitor bank characteristic. The reactive power margin from the operating point to the flat part of the *V–Q* curve is about 700 MVAr.

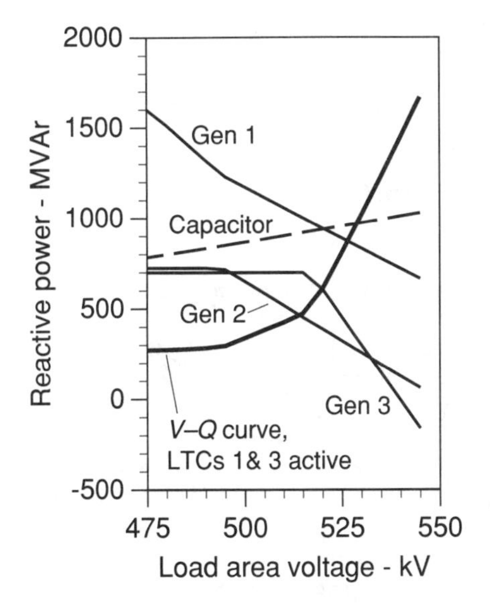

Fig. 6-9. *V–Q* curve and generator reactive power outputs for outage of one line.

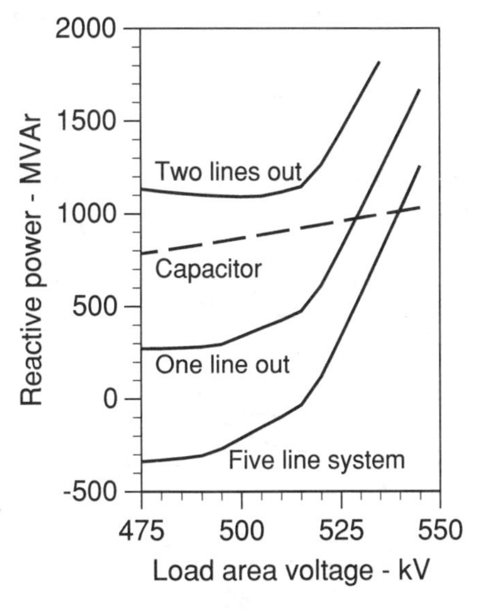

Fig. 6-10. *V–Q* curves showing effect of line outages.

V–Q curves—effect of line outages. Figure 6-10 shows *V–Q* curves for no outage, outage of one line (reference case), and outage of two lines. In all three cases, LTC transformers 1 and 3 are controlling voltage. The case with outage of two lines is unstable since there is no intersection of the capacitor bank curve and the *V–Q* curve.

V–Q curves—load representation. Figure 6-11 compares three load representations:

1. Reference case with LTC 1 and 3 in active control.
2. Prior to LTC and area interchange control. The residential/commercial bus is voltage sensitive (half resistive).
3. The resistive load at the residential/commercial bus is controlled to be constant (e.g., thermostatic control). All load is constant.

Because 75% of the load is constant *without* LTC or thermostatic control, there is not a great difference between the three cases.

With constant loads, the reactive support for the portion of the curve to the left of the critical voltage (bottom of curve) is interesting. Although not shown, the 500-kV transmission system supplies an increasing amount of reactive power as voltage is lowered. The subtransmission reactive power losses and the reduction of capacitor bank output are increasing faster, however, than the transmission system supply. This causes the curve to turn upward at low scheduled voltages.

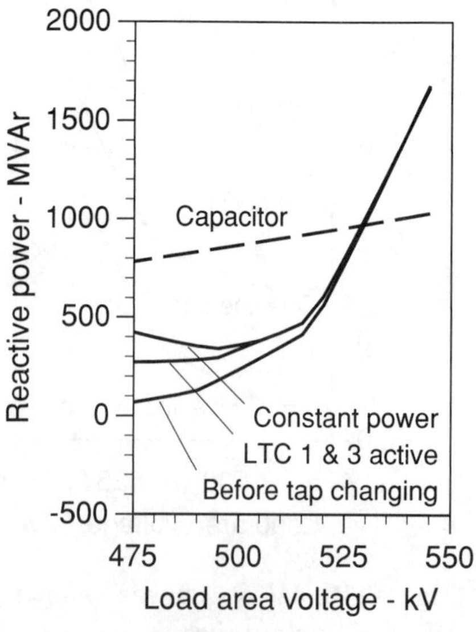

Fig. 6-11. *V–Q* curves showing effect of load control.

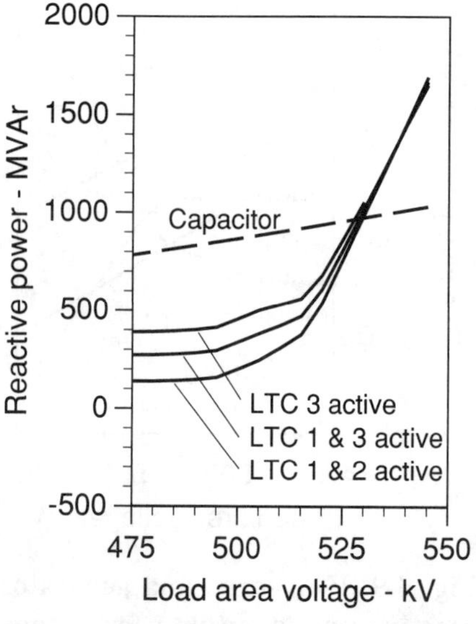

Fig. 6-12. *V–Q* curves showing effect of LTC transformers.

V–Q curves—LTC transformer control. Figure 6-12 compares different combinations of LTC transformer control. Voltage stability is improved if LTC 2 is active rather than LTC 3. LTC 2 control supports the 300 MVAr shunt capacitor bank output and reduces subtransmission losses. Voltage stability is degraded with LTC 1 inactive. This is because the industrial load shunt compensation is not supported (see Example 4-7).

V–Q curves—sending end high side voltage control. Figure 6-13 shows the improvement of voltage stability if Gen 1 and Gen 2 regulate high side voltage rather than terminal voltage (reference case). Improvement is quite significant over the entire range of scheduled voltages. This is a cost-effective method of improving voltage stability.

V–Q curves—effect of subtransmission representation. In large scale simulations, the subtransmission/distribution system is usually not represented. For constant power loads, Figure 6-14 shows the optimistic results from neglecting the subtransmission and distribution equivalent. The case with subtransmission system is the constant power case of Figure 6-11. The case without subtransmission has the power flow at the secondary of the 500/115-kV transformer converted to a constant load.

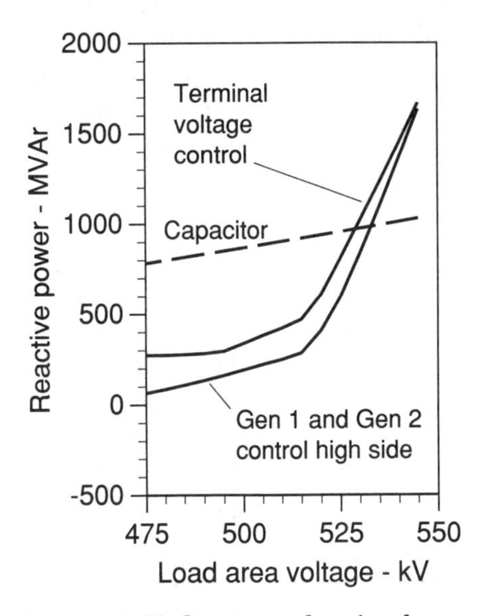

Fig. 6-13. *V–Q* curves showing benefit of high side voltage control.

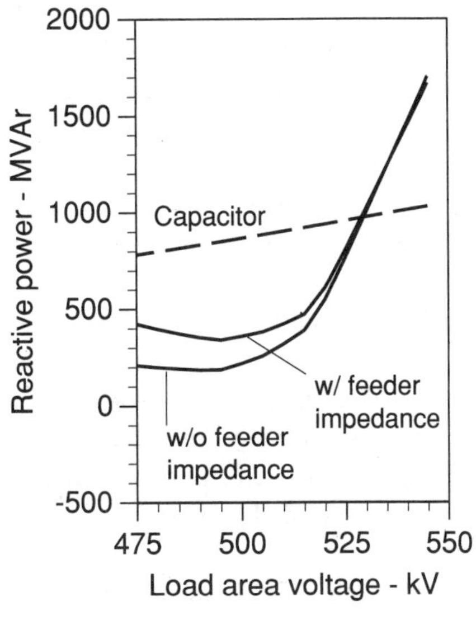

Fig. 6-14. *V–Q* curves showing optimistic results by neglecting the equivalent for subtransmission and distribution systems.

6.4 Equivalent System 2: Dynamic Simulation

We now simulate longer-term dynamics using the EPRI ETMSP 3.0 program. As described in Appendix E, the industrial load is represented by two equivalent motors. One-half of the residential/commercial load is represented by an equivalent motor. The other half is resistive with thermostatic control, i.e., constant energy load. A rather short time constant of sixty seconds is arbitrary used for this load.

Referring to Figure 6-7, LTC 3 controls the voltage at the residential load. The LTC model is shown on Figure 4-28. The delay time is thirty seconds and the mechanism time is five seconds. Discrete tap steps of 5/8% each are modeled. The bandwidth is ±0.00833 per unit corresponding to 2 volts on a 120 volt base. All other transformers have fixed tap.

Again referring to Figure 6-7, Gen 3 has an overexcitation limiter (field current limiter). The field current limit is adjusted to a value (2.17 per unit) that results in voltage instability and voltage collapse. The other generators do not have overexcitation limiting represented.

Figures 6-15 to 6-18 show simulation results for an outage of one 500-kV line at t = 5 seconds. Figure 6-15 show bus voltages in the load area. Synchronizing swings damp rapidly and the voltages are nearly constant from about t = 12 seconds until t = 35 seconds. Voltages are, however, decaying slightly because of resistive load added by thermostatic control. At t = 35 seconds tap changing begins at LTC 3. Tap changing raises the residential/commercial load voltage but lowers the other voltages. After five taps, the residential/commercial load voltage is within the bandwidth.

The system is now quiescent for about the next fifty seconds. *Then something happens.* Voltages decay and LTC 3 taps seven more times until it reaches its boost limit. Voltages collapse.

Figure 6-16 shows the cause of the voltage collapse. Gen 3 field current is slightly above the field current limiter setting for about fifty-five seconds, but then is limited at t = 110 seconds. The resulting reduction in reactive power of Gen 3 lowers the residential/commercial load voltage, causing more tap changing. Once LTC 3 is at its limit, load is added by thermostatic control. The required additional reactive power must come from the remote generators and the transmission system. This is not effective and instability results.

Figure 6-17 shows the corresponding reactive power outputs of the generators. Field current limiting reduces Gen 3 reactive power by about 55 MVAr.

Figure 6-18 show Gen 2 and Gen 3 rotor angles relative to Gen 1. Synchronous stability is not a problem. In fact, as the load is collapsing, the Gen 3 angle is moving toward the sending-end generators.

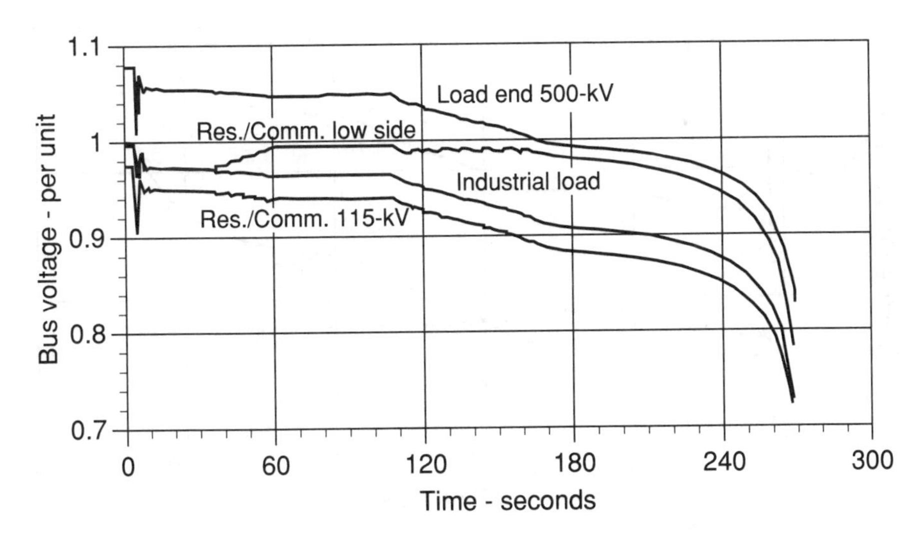

Fig. 6-15. Load area voltages for outage of one 500-kV transmission line.

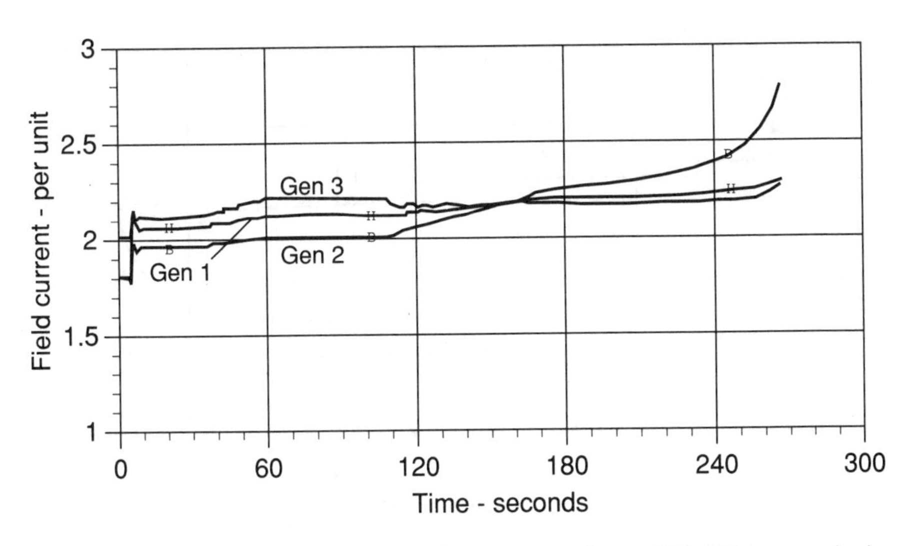

Fig. 6-16. Generator field currents for outage of one 500-kV transmission line. Field current limiting on Gen 3.

Sensitivity analysis shows that generator voltage regulator line drop compensation at Gen 1 and Gen 2 provides considerable stability improvement. This confirms the static results shown on Figure 6-13. Ideally, the line drop compensation should be adjusted so that all generators reach current limits at the same time.

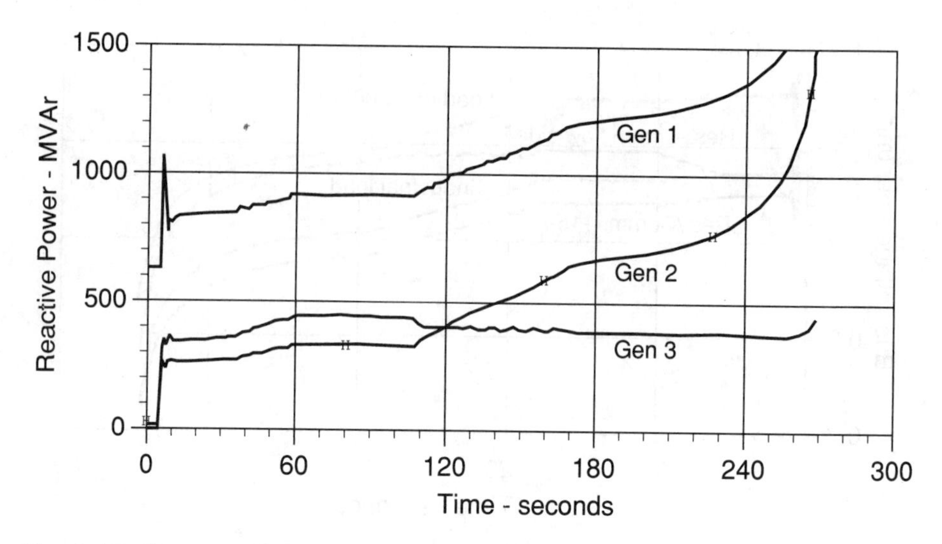

Fig. 6-17. Generator reactive power for outage of one 500-kV line. Field current limiting on Gen 3 at 110 seconds.

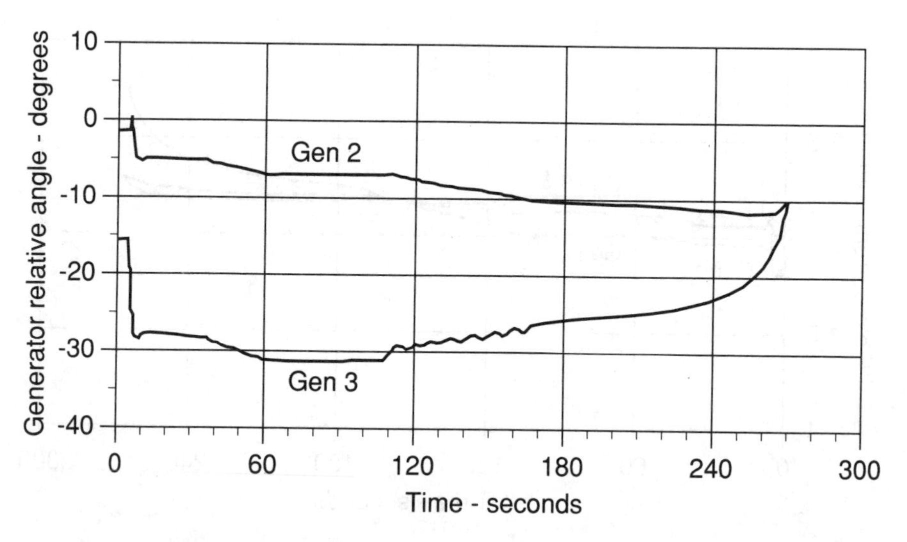

Fig. 6-18. Generator angles relative to Gen 1 for outage of one 500-kV line.

For a similar power system model, references 6 and 7 provide additional results.

References

1. C. W. Taylor, "Concepts of Undervoltage Load Shedding for Voltage Stability," *IEEE Transactions on Power Delivery*, Vol. 7, No. 2, pp. 480–488, April 1992.

2. C. V. Thio and J. B. Davies, "New Synchronous Condensers for the Nelson River HVDC System—Planning Requirements and Specification," *IEEE Transactions on Power Delivery*, Vol. 6, No. 2, pp. 922–928, April 1991.
3. A. E. Hammad and M. Z. El-Sadek, "Prevention of Transient Voltage Instabilities due to Induction Motor Loads by Static VAR Compensators," *IEEE Transactions on Power Systems*, Vol. 4, No. 3, pp. 1182–1190, August 1989.
4. A. Edris, "Controllable Var Compensator: a Potential Solution to Loadability Problem of Low Capacity Power Systems," *IEEE Transactions on Power Systems*, Vol. PWRS-2, No. 3, pp. 561–567, August 1987.
5. R. L. Hauth, S. A. Miske, Jr., F. Nozari, "The Role and Benefit of Static Var Systems in High Voltage Power System Applications," *IEEE Transactions on Power Apparatus and Systems*, Vol. PAS-101, No. 10, pp. 3761–3770, October 1982.
6. G. K. Morison, B. Gao, and P. Kundur, "Voltage Stability Analysis Using Static and Dynamic Approaches," paper 92 SM 590-0 PWRS, presented at the 1992 IEEE/PES Summer Meeting, Seattle, Washington, July 12–16, 1992.
7. CIGRÉ Task Force 38-02-10, *Modelling of Voltage Collapse Including Dynamic Phenomena*, 1993.

7

Voltage Stability of a Large System

In learning the sciences, examples are of more use than precepts.
Sir Isaac Newton

In this chapter we describe approaches by a group of utilities to solve voltage stability challenges in the major load areas of the Pacific Northwest. Interrelated voltage stability problems exist from Vancouver, British Columbia to the Puget Sound (Seattle–Tacoma) area to the Portland, Oregon area. Although we focus on the Puget Sound area, reference 1 describes approaches by B. C. Hydro to analyze voltage stability in the Vancouver load area. B. C. Hydro is removing voltage stability limitations by series capacitor and static var compensator additions.

The Pacific Northwest load areas are winter peaking and, because of large amounts of electric space heating, the loads are weather sensitive. Voltage instability is most likely during periods of cold or abnormally-cold weather, coupled with outages of key 500-kV transmission lines or major generating units.

Following discovery of serious voltage stability problems in the rapidly growing Puget Sound area, an intense effort was undertaken by Bonneville Power Administration and other area utilities to find solutions. Since solutions could involve new 500-kV transmission through the environmentally-sensitive Cascade Mountains, all possible alternatives were explored [2]. Alternatives to new transmission included reactive power compensation, load-area generation, and additional energy conservation and load management measures. We concentrate on widely applicable electric power engineering issues such as study methods, system testing, transmission system design, reactive power compensation, and power system controls.

Voltage stability problems in the Puget Sound area for the extreme contingency of loss of all transmission through a mountain pass (perhaps due to an avalanche) were known in the 1970s. Dynamic simulations showed that voltage would collapse prior to islanding and frequency collapse. Problems for single contingencies were discovered in 1987 for near-future operating conditions. Intense study of the problem started in 1988. Referring to Chapter 2, the delayed recognition of the problem may be explained by Example 2-1 which describes development of voltage stability problems in mature power systems.

The voltage stability challenge came at a time of very active industry work to understand the many aspects of the phenomena, and at a time when specialized study tools and methodology were not well developed.

7.1 System Description

The Puget Sound area peak winter load is around 11,000 MW. Peak winter loads are forecast for an annual minimum temperature with one in two years probability. During extremely cold weather (one in twenty years probability) loads are about 1000 MW higher. For Seattle, the one in two year minimum temperature is -8°C (17°F), and the one in twenty year minimum is -17°C (2°F). Load growth is 200–400 MW/year.

The main source of generation, around 8000 MW, is Columbia River hydroelectric generation on the east side of the Cascade Mountains in Washington state.

Figure 7-1 shows major transmission facilities. The cross-Cascade Mountain transmission consists of five 500-kV lines, plus several lower voltage lines. The northern corridor (mountain pass) consists of one 500-kV line (Chief Joseph–Monroe) and two 345-kV circuits. The southern corridor includes four 500-kV lines terminating at the Raver switchyard approximately 40 km southeast of Seattle. Two of the southern lines are double-circuit and series compensated (Grand Coulee–Raver 500-kV lines). A middle corridor contains one 345-kV line. The Puget Sound area is connected by north–south 500-kV transmission to the Vancouver area and to the Portland area. The major power plant in the Puget Sound area is the Centralia coal-fired plant (2 x 811-MVA units) in the extreme south. The 1280-MVA Trojan nuclear unit is still farther south, closer to Portland.

The major transmission contingencies that threaten voltage stability are outages of the Chief Joseph–Monroe 500-kV line and of the high capacity Grand Coulee–Raver double-circuit 500-kV lines.

The generation outages that threaten voltage stability are outage of the entire Centralia plant or (more probably) outage of one unit with the other unit off line, and outage of the Trojan nuclear plant.

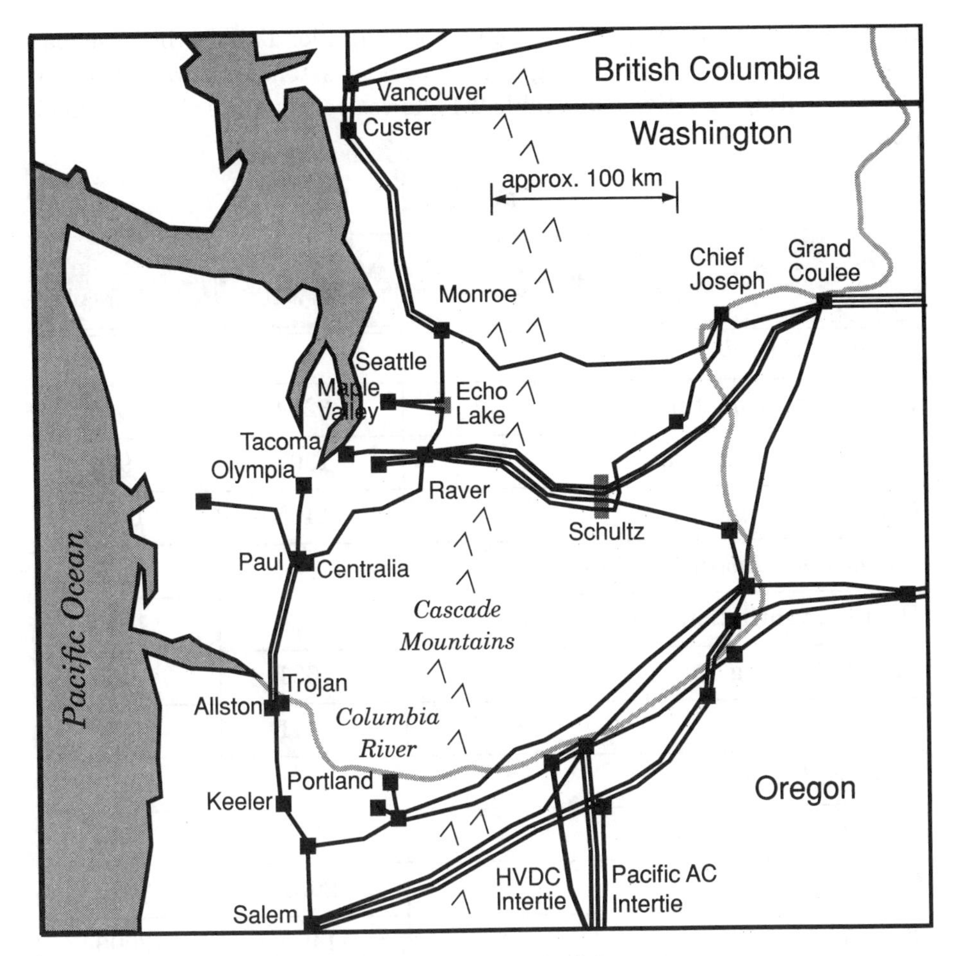

Fig. 7-1. Pacific Northwest 500-kV transmission system.

7.2 Load Modeling and Testing

Superbase case and load modeling. Puget Sound area utilities developed a "superbase" power flow base case with greatly expanded representation of the subtransmission system. Western Washington and western Oregon loads are represented at over 700 busses; the loads are regulated by LTC transformers. The entire western interconnection is represented. The total number of busses is about 5000, approximately half of which are in the British Columbia, western Washington, and western Oregon areas.

A large effort was made to improve load models for power flow and dynamic programs. The EPRI LOADSYN program described in Chapter 4 was used. Utility-derived load compositions were used in place of LOADSYN default data. The average load bus voltage sensitivity (prior to control action) is about $P \propto V^{1.3}$. The incremental load above the base load (used for example in producing a P–V curve) is temperature-dependent electric heating resistive load. Table 7-1 shows overall winter load composition [2].

Table 7-1

Winter Peak Loads	Normal peak (MW)	Extreme peak (MW)
Residential loads		
Electric space heating	3244	4306
Electric water heating	1307	974
Other	1465	1403
Commercial / institutional loads		
Heating, ventilation & air conditioning	1526	1859
Electric lighting	681	755
Other	341	374
Industrial loads		
Aluminum plants	755	759
Other	2081	2070
Total load	11400	12500

A key question in the voltage stability analysis is load modeling. Should loads simply be represented as constant power as seen from transmission or subtransmission busses? The load modeling question is complex: for example, during heavy load, tap changers may be near boost limits with limited range for voltage and load regulation following an outage. Some utilities (Seattle City Light in the Puget Sound area) don't use tap changing equipment for voltage regulation.

The superbase case was developed essentially for sensitivity analysis of load modeling. For development of transmission plans for later years, normal representation (less subtransmission system) may be used, with the assumption of constant power loads.

Load testing. Several methods were used to understand load behavior. During abnormally cold weather in February 1989 and December 1990, tap changer tap positions were noted. On average, a regulating margin remained to correct about a 6% voltage drop. System tests were also performed to measure load response following a voltage drop. Figure 4-34 shows a test result at Port Angeles on the Olympic Peninsula.

During January and February 1991, two large-scale tests were performed to observe load behavior for the entire Puget Sound area. Under controlled conditions, the two Grand Coulee–Raver 500-kV lines were opened sequentially to drop system voltage. The lines are 280 km long. Also, a 180-MVAr, 500-kV shunt reactor was energized at Raver. For the test on February 26, the switching dropped Raver voltage by 22 kV (-4.1%). The voltage dropped another 4 kV during the next two minutes as voltage-sensitive load was restored. The weather was mild for this test.

Load tap changer (LTC) transformer responses were monitored at many substations. Figures 7-2 and 7-3 show measurements at a suburban 115-kV substation just south of Seattle. The active and reactive power measurements are for a 115/12.5-kV LTC transformer (12/16/20 MVA rating). The load is estimated at 73% residential and 27% commercial. From the LOADSYN program (for heavy load, cold weather conditions), $\Delta P/\Delta V = 1.64$ per unit/per unit and $\Delta Q/\Delta V = 2.33$ per unit/per unit.

Figure 7-2 shows response from one minute before the outages until fourteen minutes after the outage. Figure 7-3 shows response from ten seconds before the outage until 120 seconds after the outage. The outages and reactor switching dropped the 115-kV voltage by about 4 kV (measured at about 70 seconds, see Figure 7-3a). The corresponding drops in load power were 0.5 MW and 0.164 MVAr. Based on the predisturbance values, the per unit voltage sensitivities are:

$$\frac{\Delta P}{\Delta V} = \frac{0.5/9.18}{4/117.2} = 1.60$$

$$\frac{\Delta Q}{\Delta V} = \frac{0.164/2.2}{4/117.2} = 2.18$$

Starting about forty seconds after the first outage, the voltage decayed another 1.2 kV as load was restored by tap changing. The voltage recovery between three and four minutes is probably due to high side voltage control at generating plants. Approximately forty seconds after the line reclosures, voltage slowly increased as tap changing reduced load.

On Figure 7-3b, the tap positions are indicated (L means lower, R means raise). The time delay setting is 30 seconds. This time delay is

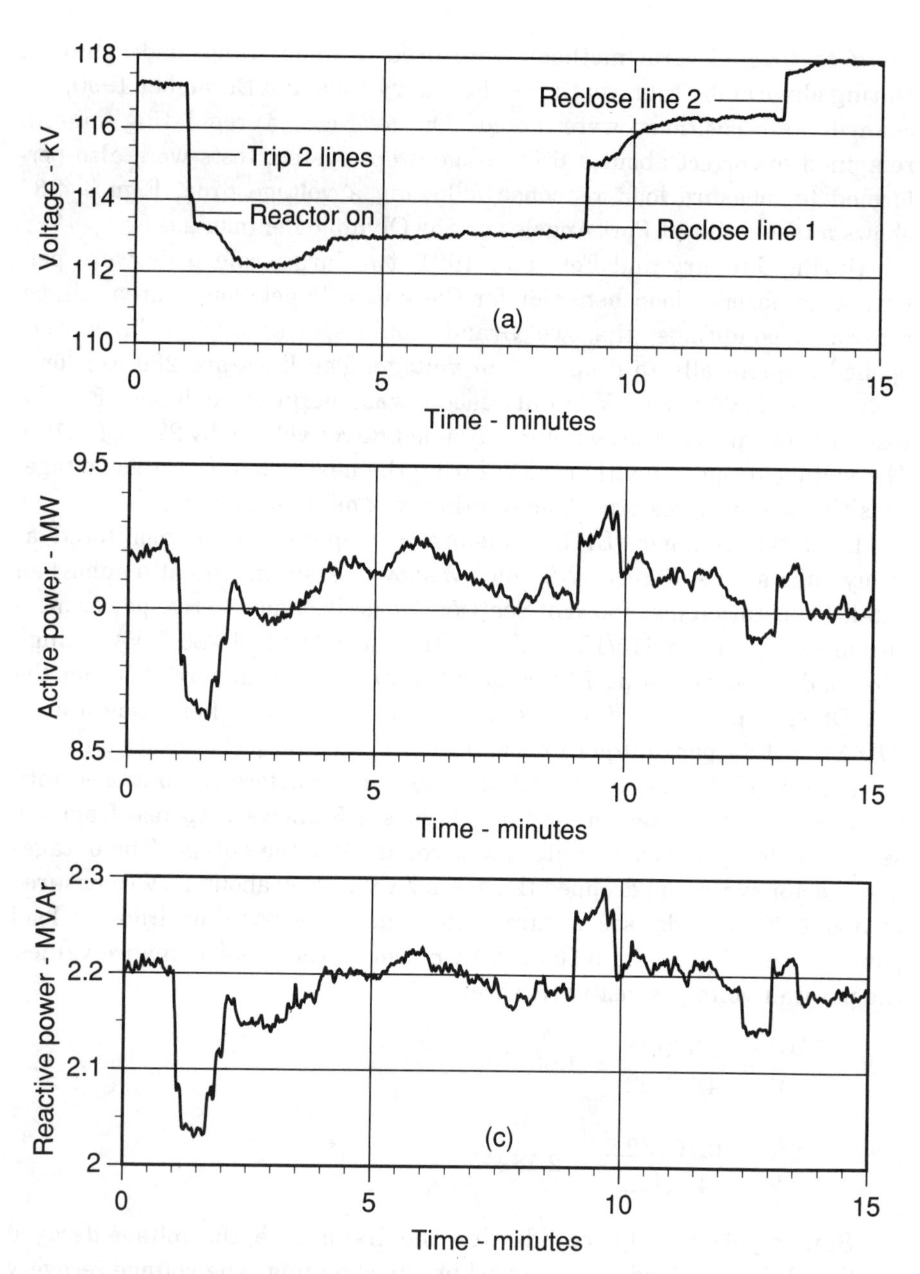

Fig. 7-2. Puget Sound area tests on February 26, 1991. Measurements at a 115-kV substation south of Seattle.

shown (we assume the timer relay is energized after the second line outage). Between 99 and 122 seconds, the transformer taps from position 4L to 2R (6 steps of 5/8% each). The mechanism time from an even numbered

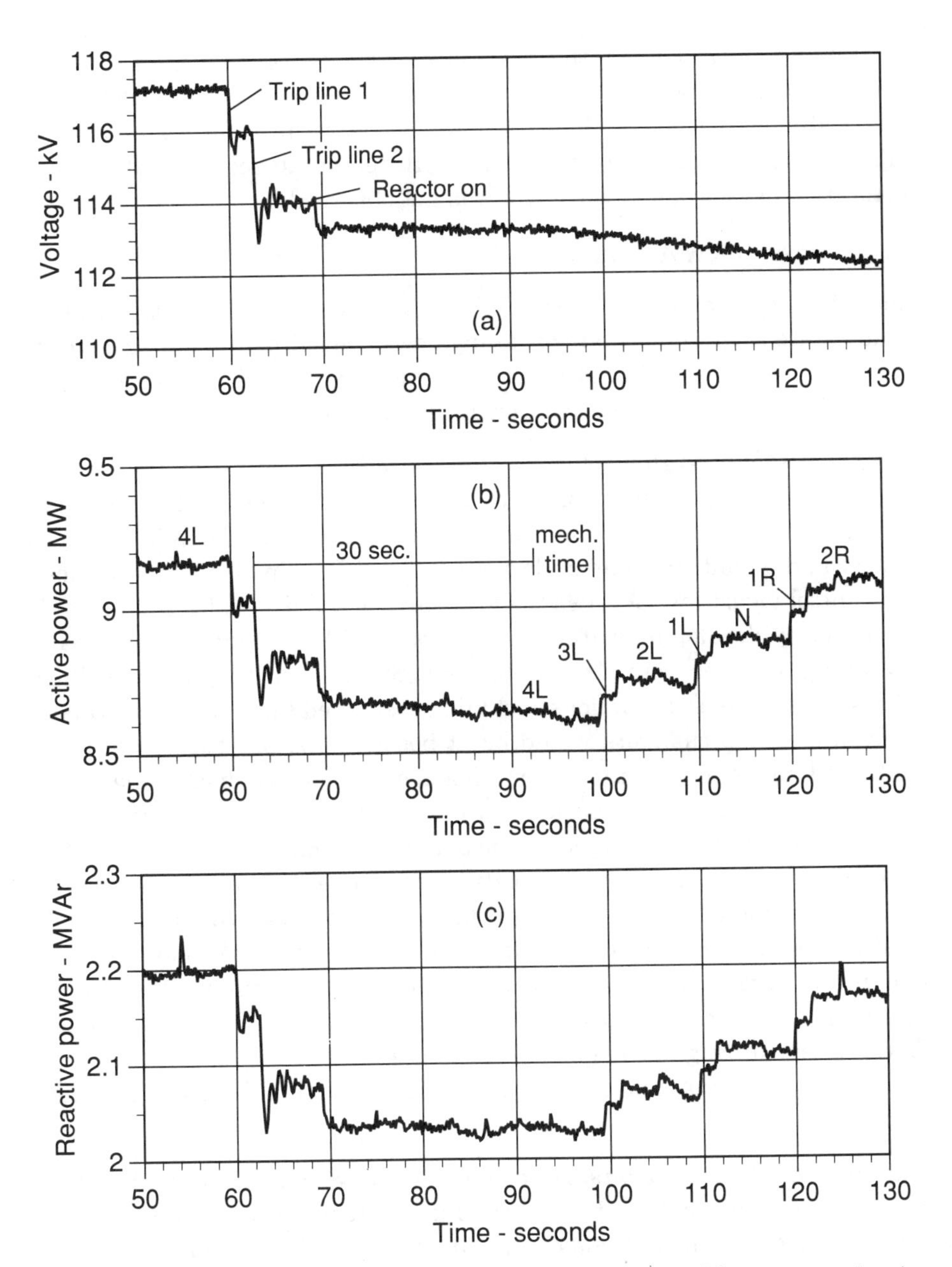

Fig. 7-3. Puget Sound area tests on February 26, 1991. Measurements at a 115-kV substation south of Seattle. Tap positions are shown on Figure 7-3b.

position to an odd numbered position is about eight seconds and the time from an odd position to an even position is about two seconds. The Westinghouse T & D book [3] describes several types of tap changer equipment.

Because over 30% of the load is electric space heating and other heating, response was tested by reducing voltage on several feeders with electrically heated homes. Results, however, were inconclusive because of mild weather during the test periods. With temperatures above 0° C, load restoration by thermostatic control was not observed.

7.3 Power Flow Analysis

Most of the system design by the Bonneville Power Administration was by steady-state power flow analysis. Many power flow program enhancements aided the voltage stability studies. In particular, the *V–Q* curve technique was automated and became the workhorse method.

Deterministic design criteria. Value-based (probabilistic) calculations showed that the joint probability of a major outage and peak load conditions is very, very low. Furthermore, the consequences of a voltage emergency could be mitigated by undervoltage load shedding [4] and other low-cost control measures. Because the loads are inherently highly voltage sensitive, slowly-developing voltage collapse would likely result in abnormally low voltages rather than a total blackout. Stable operation at low voltage allows time for manual actions including load shedding. With abnormally low voltages, tap changers would be at boost limits, but thermostatically controlled and other constant energy loads may be slowly increasing.

Nevertheless, utilities decided to use traditional deterministic design criteria. The system is designed to be stable with margin for a first contingency during abnormally heavy loads (once in twenty years cold weather) and for a second contingency for normal peak loads (once in two years cold weather [5]. The system is also designed to serve the load demanded at nominal voltage, and all voltages are required to be within specified limits. The most limiting condition is usually the double-circuit Grand Coulee–Raver outage during normal peak load conditions.

Design (planning) and operating criteria are further discussed in Chapter 9.

Voltage stability solutions. Based on power flow simulation (mainly *V–Q* curves as described below), two new 500-kV switchyards and many reactive power compensation additions were identified. This postpones new line construction. For initial reinforcements, the following equipment is added (refer to Figure 7-1).

- Three 152-MVAr, 230-kV mechanically switched shunt capacitor banks (at three Puget Sound area substations).
- One 316-MVAr, 500-kV mechanically switched capacitor bank at Monroe.

- One ±300 MVAr static var system connected to the Maple Valley 230-kV bus. The Maple Valley 230-kV bus is connected to the 500-kV bus by a 1600-MVA transformer. A cross-Cascade Mountain 345-kV line also terminates at this bus.
- New 500-kV switchyard at Echo Lake. This reduces the electrical distance between Raver and Monroe caused by looping the Raver–Monroe line into Maple Valley.
- Two 316-MVAr, 500-kV mechanically switched capacitor banks at Echo Lake.
- Line drop compensation (50%) at the Centralia and Grand Coulee power plants, and fast high side voltage control at the Grand Coulee, Chief Joseph, Bonneville, The Dalles, and John Day hydroelectric power plants along the Columbia River.
- Other 500-kV MSCs and a static var system in the Portland area.

A large new 500-kV switchyard (Schultz) is added along the southern corridor about 130 km east of Raver. All four 500-kV lines will be looped into the switchyard. Outage of the double circuit line (Schultz–Raver or Schultz–Grand Coulee) becomes much less severe because of the intermediate switchyard. Series compensation is planned at Schultz on the four Raver lines (initially only on the high capacity double circuit line).

In a ten year time frame, a new high capacity transmission line may be needed. This will probably be a double-circuit, series compensated 500-kV circuit in either the northern or middle corridor. The new line reduces peak losses by around 100 MW, and relieves severe overloads during outages. Several other transmission line and transformer additions are also required in this time frame.

V–Q curves. *V–Q* curves computed for the Raver 500-kV bus were used to rapidly assess Puget Sound area bulk system voltage security. At this node, four heavily loaded 500-kV lines from generation areas to the east terminate. A fifth 500-kV line runs to the north and a sixth 500-kV line runs to the south. Three more 500-kV lines serve Tacoma area loads. There are also three 500-kV shunt capacitor banks at Raver totaling over 1000 MVAr.

For conditions prior to voltage stability related reinforcements, Figure 7-4 shows the Raver *V–Q* curve for three load conditions following the Grand Coulee–Raver double circuit outage. The three cases are:

1. Voltage sensitive loads—this represents a snapshot in time prior to tap changing, perhaps 30 seconds following the outage.
2. Voltage sensitive loads with discontinuous tap changing—for the lower voltages at Raver, tap changers are at boost limits and the loads are voltage sensitive.

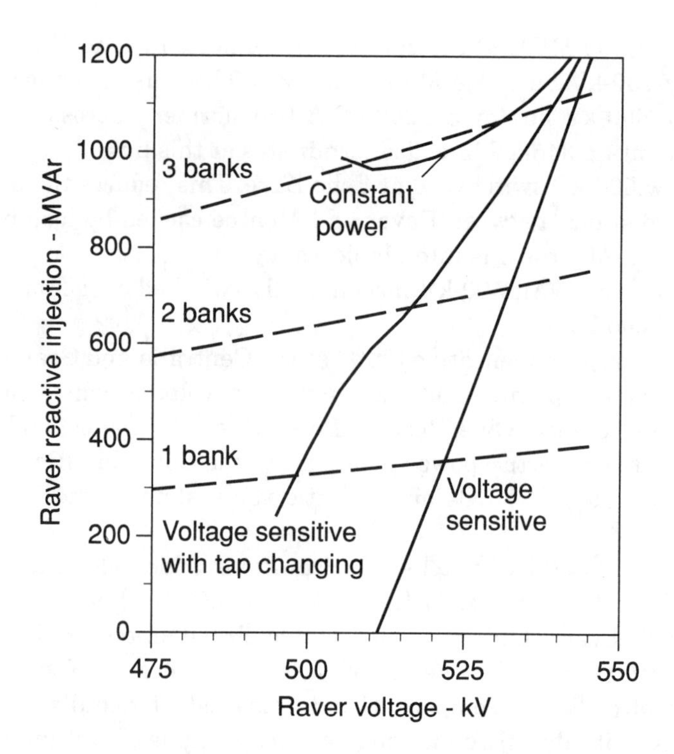

Fig. 7-4. *V–Q* curve for the Raver 500-kV bus. Outage of Grand Coulee–Raver double circuit 500-kV line. Superbase case with reduced load levels.

3. Conventional constant power load representation. This approximates a point in time when thermostats and manual actions have controlled constant energy type loads. Remaining voltage sensitive loads such as lighting are treated as constant.

Figure 7-4 also shows the *V–Q* characteristics of the three Raver shunt capacitor banks. The intersections of the capacitor bank and system characteristics are possible operating points. For the constant power load assumption there is very little stability margin.

Figure 7-5 shows the *V–Q* curve with load growth and with the network reinforcements except for the Schultz switchyard. Constant power loads are assumed. Note that the critical voltage for the Grand Coulee–Raver outage is 525-kV. The most severe case is the Grand Coulee–Raver outage during normal peak loads.

Figure 7-6 shows the *V–Q* curve for the ten year time frame with the Schultz switchyard. The most severe disturbance is now the Chief Joseph–Monroe 500-kV outage with extra heavy load levels.

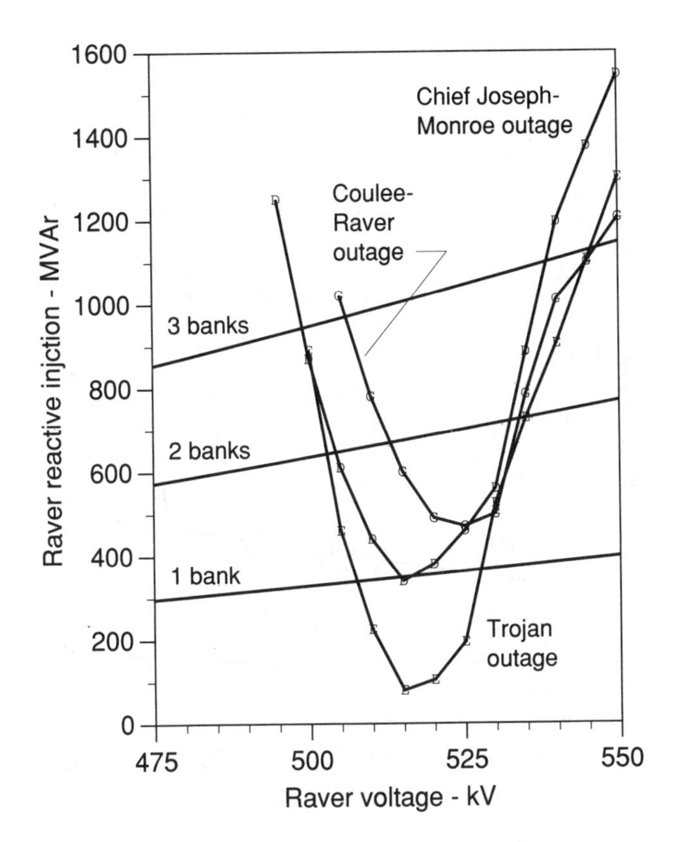

Fig. 7-5. Raver V–Q curve with network reinforcements. Grand Coulee–Raver double circuit outage for normal peak load conditions. Chief Joseph–Monroe and Trojan outage for extra heavy load conditions. Constant power loads.

For the cases shown on Figures 7-4 and 7-5, a 500 MVAr reactive margin from the operating point to the bottom of the constant power V–Q system characteristic is required. This margin allows for unavailability of a Raver, Echo Lake, or Monroe 316 MVAr shunt capacitor bank, or the Maple Valley SVC. The margin also allows for other uncertainties such as higher load levels. For these highly stressed outage conditions, two or three megavar of reactive support might be required for each additional megawatt of load.

7.4 Dynamic Performance Including Undervoltage Load Shedding

While system additions were based mainly on V–Q curves and constant power loads, dynamic simulation provides a more realistic picture of system voltage stability. V–Q curves artificially stress a key transmission

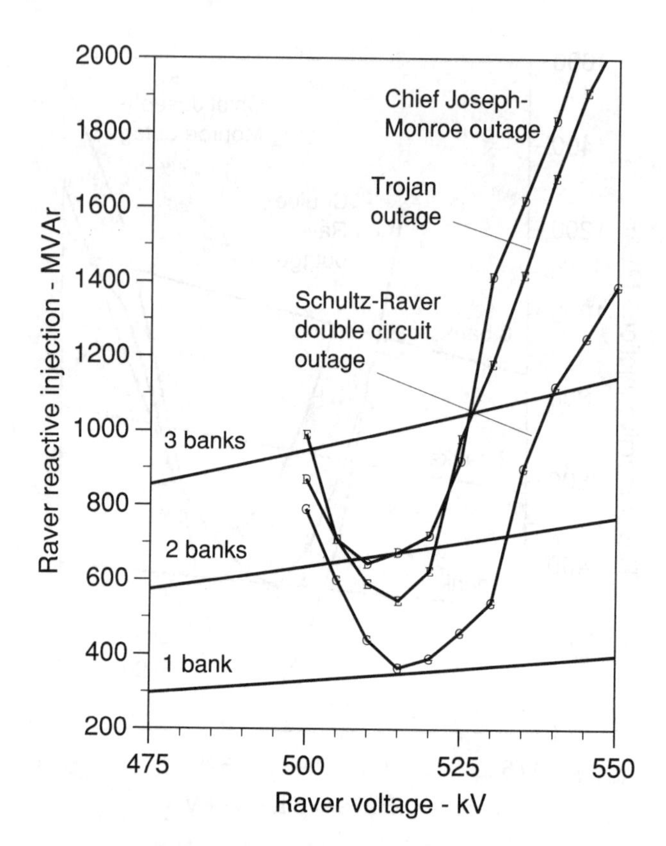

Fig. 7-6. Raver *V–Q* curve for ten year time frame. Schultz–Raver double circuit outage for normal peak load conditions. Chief Joseph–Monroe and Trojan outage for extra heavy load conditions.

bus and do not simulate actual operating points. *V–Q* curves at Raver provide only a crude approximation of the dynamic performance and voltage security of the geographically-large Puget Sound area. For wintertime conditions with voltage sensitive loads, Figure 2-14 shows that assuming constant power loads is not realistic unless there is a requirement to serve the load demanded at nominal voltage.

Reference 4 reports initial dynamic simulation results, including undervoltage load shedding, using a conventional transient stability program. We now describe results using the EPRI ETMSP 3.0 program for extended dynamic analysis.

The ETMSP simulation program includes models for underload tap changing, for thermostatically-controlled heating loads, and for generator field current limiting. A superbase power flow case is used. The simulated system comprises 4960 busses, 8898 branches, 411 generators, 573 induction motors, 719 under-load tap changing transformers (discrete taps with limits and deadband), 729 thermostatically controlled loads, and about fifty undervoltage load shedding relays. To save computer time, the thermostatically-controlled loads are assumed to have a time constant of only sixty seconds; a 20% increase in conductance is allowed.

Seven major generators have overexcitation limiters modeled. The Centralia Power Plant and the Grand Coulee Third Power Plant have line drop compensation represented. Both power plants connect to the 500-kV network, and compensation in both cases is 50% of the step-up transformers.

The 15% of area load with undervoltage load shedding represents about 1800 MW of area load [4]. The Puget Sound area load shedding program, implemented in December 1991, is as follows:

1. 5% of area load is shed at a voltage 10% below lowest normal voltage. The time delay is 3.5 seconds.
2. 5% of area load is shed at a voltage 8% below lowest normal voltage. The time delay is 5 seconds.
3. 5% of area load is shed at a voltage 8% below lowest normal voltage. The time delay is 8 seconds.

Extra-heavy loads are represented with the system additions described in the previous section not included.

Figures 7-7, 7-8, and 7-9 show large-scale time domain simulation results of two minute duration. The disturbance is outage of the Grand Coulee–Raver double circuit 500-kV line loaded at 2866 MW. This severe disturbance is outside the deterministic planning criteria for the extra-heavy load conditions represented.

Figure 7-7 shows selected 500-kV voltage. After synchronizing swings have damped out, there is some voltage decay prior to tap changing because of increase in thermostatically-controlled resistive load. Because of the very severe disturbance, there is also some undervoltage load shedding prior to tap changing—which starts after thirty seconds.

At the end of the simulation, some 500-kV voltages are 0.96 per unit or 480 kV which is about 10% below normal. Most under-load tap changing transformers reach their boost limits, and eleven of the thermostatically controlled loads reach their maximum value.

Figure 7-8 shows selected unregulated 115-kV voltages. Regulated voltages on the load side of LTC transformers are generally boosted by 10%.

Figure 7-9 shows field current at important generators. After time delay, field current limiting is enforced at the Trojan and Grand Coulee generators. Centralia is near its maximum field current, but load shedding has prevented significant overload during the simulation duration.

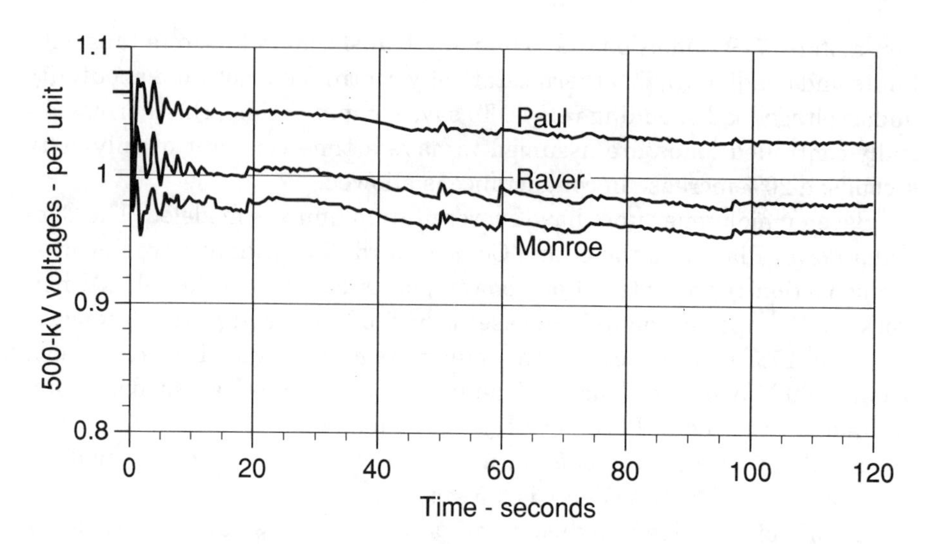

Fig. 7-7. Raver, Monroe, and Paul 500-kV bus voltages for outage of the Grand Coulee–Raver double circuit 500-kV line loaded at 2866 MW. The drops in voltage are due to tap changing and the rises are due to undervoltage load shedding.

The amount of undervoltage load shedding is 686 MW and 187 MVAr, with shedding of twelve blocks of load. Although voltages are abnormal, the system is *stable* and *secure*. About 1100 MW of undervoltage load shedding remain in reserve to protect against unexpected transmission line or power plant relaying caused by the abnormal conditions. In the simulation world, the reserve undervoltage load shedding "protects" against modelling errors. One or two 500-kV shunt capacitor bank additions would improve system performance.

Sensitivity cases show that generator line drop compensation installed at the Centralia power plant and the Grand Coulee third power plant significantly reduces the amount of undervoltage load shedding. For the outage described, the line drop compensation reduces load shedding from 1040 MW to 686 MW. Final voltage levels are similar for both cases.

A case without load shedding also is stable, but probably not secure. Voltages at the 500-kV level are about 20% below normal, and major generators have field current limited with stator current overloads. The system is stable because of the high voltage sensitivity of electric heating and other load once tap changers are at boost limit, and all thermostatically-controlled electric heating load is connected.

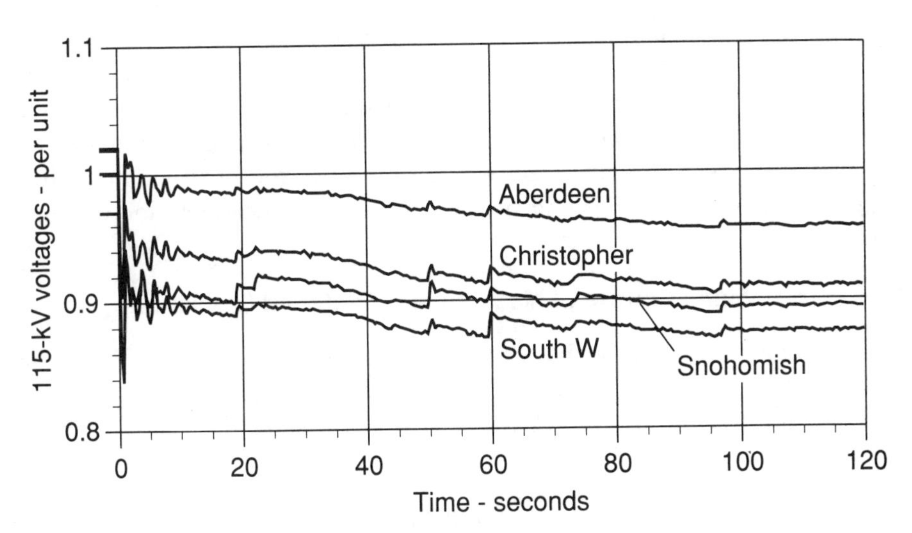

Fig. 7-8. Selected 115-kV bus voltages for outage of the Grand Coulee–Raver double circuit 500-kV line loaded at 2866 MW. The drops in voltage are due to tap changing and the rises are due to undervoltage load shedding.

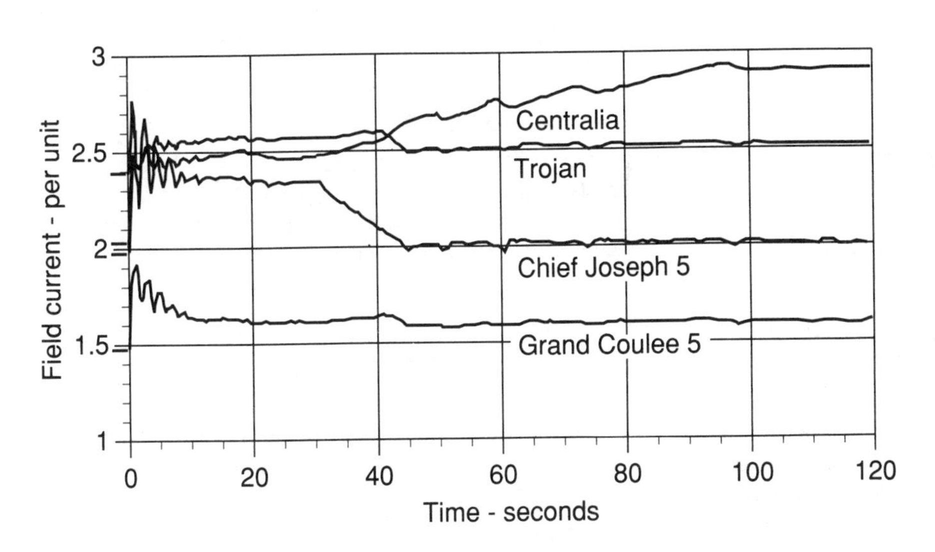

Fig. 7-9. Field current at Centralia, Trojan, Chief Joseph and Grand Coulee for outage of the Grand Coulee–Raver double circuit 500-kV line loaded at 2866 MW. The Grand Coulee and Chief Joseph generation are units connected to the 500-kV system. Field current limiting is enforced at Trojan and Chief Joseph.

The worst case within the deterministic planning criteria, outage of the Chief Joseph to Monroe 500-kV line loaded at 1594 MW, is stable. Load shedding is 196 MW and 50 MVAr. Voltages at the 500-kV level are about 8% below normal. Although several additional shunt capacitor banks are effective (i.e., cost effective) and would eliminate the need for load shedding, the more expensive additions do not seem necessary for the near term.

7.5 Automatic Control of Mechanically Switched Capacitors

The Puget Sound area mechanically switched capacitor banks (MSCs) are controlled by dispatchers, and by local voltage-based controllers. For the Raver 500-kV Banks 2 and 3, Tables 7-2 and 7-3 show the voltage controller settings.

Table 7-2

	Cut in #1 kV	Delay seconds	Cut in #2 kV	Delay seconds
Raver Bank 2	520	10	510	10
Raver Bank 3	520	15	510	6

Table 7-3

	Cut out #1 kV	Delay seconds	Cut out #2 kV	Delay seconds
Raver Bank 2	555	20	565	10
Raver Bank 3	555	10	565	5

These settings have been used for several years without undesirable operations. Generally, the 520 kV setting is 10–20 kV or 2–4% below predisturbance bus voltage and, following a major disturbance, they may not be energized until significant load has been restored by tap changers. More sensitive automatic control methods are desirable to energize capacitor banks at a higher voltage—before tap changing causes significant voltage decay. We describe a new method for centralized control of capacitor banks.

Premise. A key to voltage security is maintaining sufficient fast-acting reactive power reserves at critical generating units and at static var compensators. Shunt capacitor banks are switched to allow near unity power factor operation of generators and SVCs during normal conditions (preventive control). Following major disturbances, capacitors banks are switched to restore the fast-acting reactive reserves (corrective control). The following pertains mainly to corrective control following large disturbances.

Voltage control strategy for voltage stability. At generating plants, transmission (high side) voltage is maintained near maximum level by line drop compensation or by automatic high side voltage control. This is local control with infrequent changes in scheduled high side voltage from the control center; fast centralized control is not necessary nor desirable.

The Maple Valley SVC controls 400 MVAr of 230-kV capacitor banks at Maple Valley. Local control of the capacitor banks by the SVC will help maintain SVC reactive power reserves following severe disturbances. Again, central control is not necessary nor desirable.

The remaining question is how to obtain fast, sensitive control of other mechanically switched capacitor banks.

Automatic control of mechanically switched capacitors. Bonneville Power Administration is considering an automatic voltage and reactive power control system for both normal and emergency conditions. The fast emergency control is based on local voltage measurement augmented by remote signals from change of reactive power output at generating plants and SVCs.

Inspired by the microprocessor-based substation voltage/reactive power controller developed by Tokyo Electric Power Company [6], Figure 7-10 shows a possible substation controller characteristic for a substation such as Monroe that has both 500-kV and 230-kV capacitor banks, and a 500/230-kV LTC autotransformer. Appropriate switching is ordered after accumulated kV-seconds outside the deadzone reaches a set value. The V_{aux} signals from a central controller biases the local voltage measurements.

For the example that follows, however, we only consider control of 500-kV MSCs based on accumulated (integrated) kV-seconds below 520 kV. The controllers will reset immediately if voltage rises above 520 kV due to capacitor bank switching at other locations, line reclosures, or other switching. The switching time delay will always be long enough to allow an automatic line reclosure attempt. Figure 7-11 shows this simpler 500-kV MSC controller. The $V_{aux\,500}$ signal comes from a central controller.

Equation 7.1 describes the central controller signal sent to the substation MSC controllers. Reactive power signals are received from the Centralia Power Plant, the Grand Coulee Third Power Plant, and the Maple Valley SVC. These signals are passed through a washout (high pass) filter to obtain reactive power change following a disturbance. The washout filter time constant is 2–5 minutes. Both the Centralia and Grand Coulee power

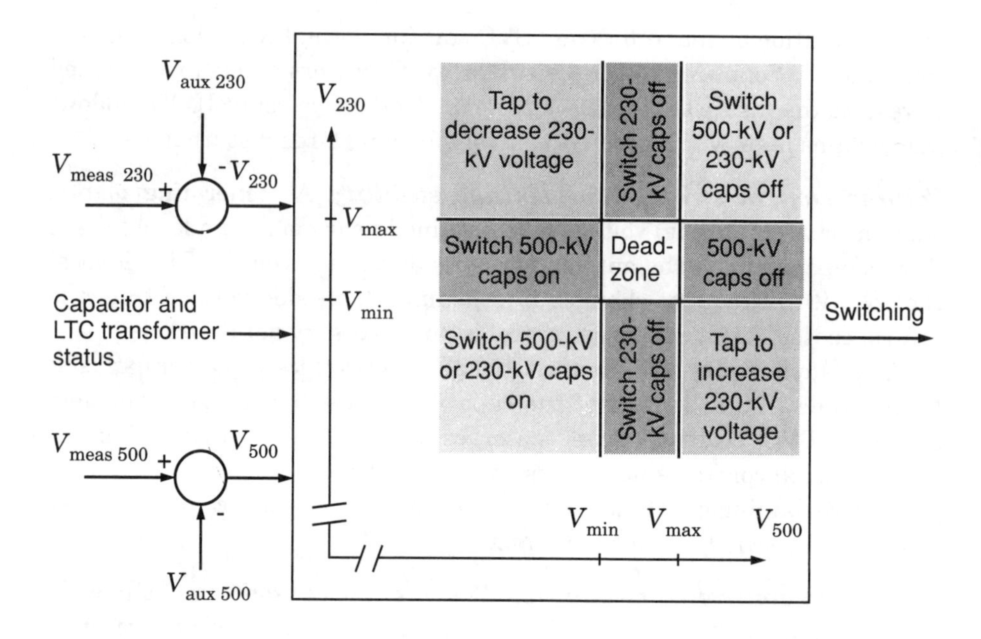

Fig. 7-10. Microprocessor-based substation voltage/reactive power controller with both 500-kV and 230-kV MSCs, and LTC autotransformer.

plants are connected to the 500-kV system, and the voltage regulators of both power plants have 50% line drop compensation. Thus the changes in reactive power of the power plants provide sensitive indications of disturbances requiring corrective action.

$$V_{aux\ 500} = K_1 \Delta Q_{Centralia} + K_2 \Delta Q_{GC5} + K_3 \Delta Q_{MV\ SVC} + U \qquad (7.1)$$

The U term in Equation 7.1 is available for voltage control during normal operation. Based on optimal power flow, expert system, or other methods, a large value of U would initiate switching.

Numerical example. The example is for a case with the Maple Valley SVC out of service—only the Centralia and Grand Coulee reactive power signals are used. Post-disturbance power flow simulation is used representing snapshots in time during the first thirty seconds following outage of the Chief Joseph–Monroe 500-kV line loaded at 1495 MW. The superbase case is used. Referring to Table 7-1, extreme peak loads are represented. Loads are represented as voltage sensitive (i.e., prior to load restoration by tap changing). The significant generation-load imbalance due to the voltage-sensitive loads are compensated for by using the "governor power flow" method. In the base case, 500-kV MSCs at Monroe and Echo Lake are

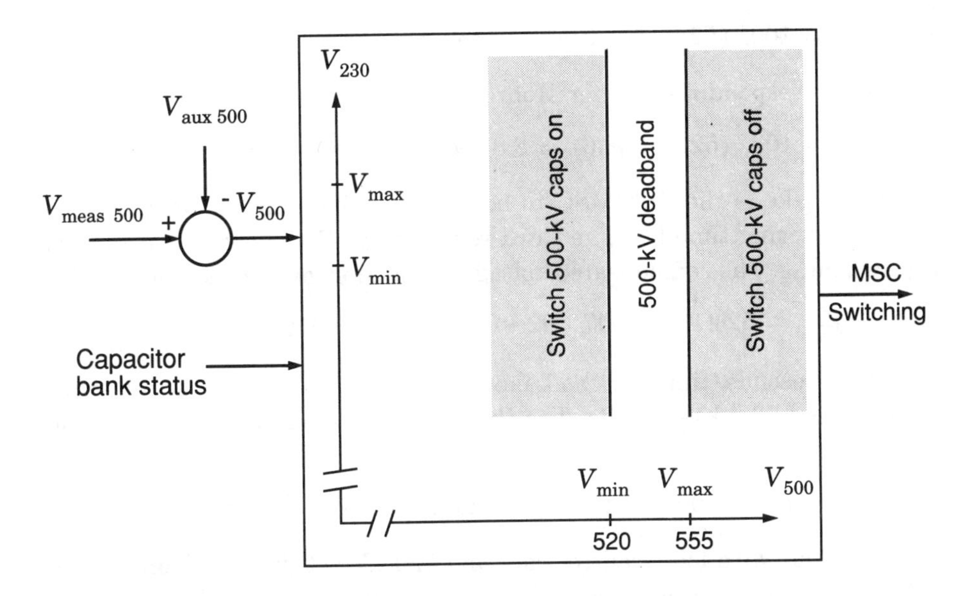

Fig. 7-11. Microprocessor-based substation voltage/reactive power controller with 500-kV MSC control only.

available for post-disturbance switching. We will not consider control of 230-kV MSCs.

Referring to Equation 7.1, the gain settings for Centralia and Grand Coulee are 0.007 kV/MVAr and 0.02 kV/MVAr, respectively (same gains used at both Monroe and Echo Lake). For both Monroe and Echo Lake MSCs, switching occurs after accumulation of 100 kV-seconds below 520 kV.

Table 7-4, rows one and two, shows the base case conditions and the effect of the outage. The outage drops Puget Sound area 500-kV voltages 2–3%. Based on Equation 7.1, and Figure 7-11, the compensated voltage of the Monroe substation controllers following the outage is:

$$V_{500} = 517.0 - (0.007 \times 156 + 0.02 \times 363) = 517 - 8.4 = 508.6 \text{ kV}$$

Note that the remote signals makes the controller input voltage 8.4 kV lower then the locally measured voltage alone. The corresponding compensated voltage at Echo Lake using the same gains is:

$$V_{500} = 514.7 - (0.07 \times 156 + 0.02 \times 363) = 506.3 \text{ kV}$$

Ignoring synchronizing oscillations and other transients, the time to accumulate 100 kV-seconds below 520 kV for Echo Lake is:

$$t = 100/(520 - 506.3) = 7.3 \text{ seconds}$$

The corresponding time for Monroe is:

$$t = 100/(520 - 508.6) = 8.8 \text{ seconds}$$

Echo Lake is the first MSC to be energized by a circuit breaker with five cycle closing time [7]. The third row of Table 7-4 shows the results of the switching. The compensated voltage of the Monroe MSC is now:

$$V_{500} = 524.1 - (0.007 \times 92 + 0.02 \times 267) = 518.1 \text{ kV}$$

At 7.3 seconds (time of Echo Lake switching), the Monroe controller has accumulated 83.2 kV-seconds. The time for the Monroe MSC energization is:

$$t = 7.3 + (100 - 83.2)/(520 - 518.1) = 16.1 \text{ seconds}$$

Note that with local control only and 520 kV setting, the Monroe controller would have reset without the remote signal.

Table 7-4

	ΔQ - MVAr		Voltages - kV		Puget Sound Load MW
	Centralia	Grand Coulee	Monroe	Echo Lake	
Base	-	-	528.1	530.0	12,212
Outage[a]	156	363	517.0	514.7	11.815
With MSC[b]	92	267	524.1	524.5	11,986
With MSC[c]	46	197	537.0	531.6	12,149

a. Chief Joseph–Monroe 500-kV line outage.
b. Chief Joseph–Monroe outage with Echo Lake MSC inserted.
c. Chief Joseph–Monroe outage with Monroe MSC also inserted.

The fourth row of Table 7-4 shows the result of the Monroe MSC switching. The Puget Sound area load is restored to within 0.5% of the predisturbance value. This indicates that voltages are mostly returned to predisturbance values, and that there would be very little tap changing at bulk power delivery LTC transformers and distribution voltage regulators. The fast switching prevents voltage decay by tap changing or thermostatic control of loads. The centralized control improves the effectiveness of the

MSCs. Allowing voltage decay and energizing MSCs at a lower voltage reduces the reactive power output of all shunt capacitor banks. For the condition studied with the centralized MSC control, the Maple Valley SVC does not appear necessary.

References

1. IEEE Committee Report, *Voltage Stability of Power Systems: Concepts, Analytical Tools, and Industry Experience*, IEEE publication 90TH0358-2-PWR, 1990.
2. Bonneville Power Administration, *Puget Sound Area Electric Reliability Plan—Final Environmental Impact Statement*, DOE/EIS - 0160, April 1992.
3. Westinghouse Electric Corporation, *Electrical Transmission and Distribution Reference Book*, East Pittsburgh, Pennsylvania, 1964.
4. C. W. Taylor, "Concepts of Undervoltage Load Shedding for Voltage Stability," *IEEE Transactions on Power Delivery*, Vol. 7, No. 2, pp. 480–488, April 1992.
5. B. L. Silverstein and D. M. Porter, "Contingency Ranking for Bulk System Reliability Criteria," *IEEE Transactions on Power Systems*, Vol. 7, No. 3, pp. 956–964, August 1992.
6. S. Koishikawa, S. Ohsaka, M. Suzuki, T. Michigami, and M. Akimoto, "Adaptive Control of Reactive Power Supply Enhancing Voltage Stability of a Bulk Power Transmission System and a New Scheme of Monitor on Voltage Security," *CIGRÉ*, paper 38/39-01, 1990.
7. B. C. Furumasu and R. M. Hasibar, "Design and Installation of 500-kV Back-to-Back Shunt Capacitor Banks," *IEEE Transactions on Power Delivery*, Vol. 7, No. 2, pp. 539–545, April 1992.

8

Voltage Stability with HVDC Links

Progress is trouble.
Charles Kettering, former chief engineer at General Motors

High voltage direct current (HVDC) links are used for extremely long distance transmission and for asynchronous interconnections. An HVDC link can be either a back-to-back rectifier/inverter link or can include long distance dc transmission. Multi-terminal HVDC links are feasible. Figure 8-1 shows suspended HVDC valves for a ±500-kV converter. Figure 8-2 shows an ac harmonic filter.

The technology has matured to the point that HVDC terminals can be connected at voltage-weak points in power systems. By voltage-weak, we mean that switching load or shunt reactive compensation causes a large voltage change. This is due to a high impedance source system, or due to heavy ac power transfers.

HVDC links may present unfavorable "load" characteristics to the power system. The dc "load" in question is usually the reactive power consumption of an HVDC inverter station providing power to a generation-short load area. An HVDC converter (rectifier or inverter) consumes reactive power equal to 50–60% of the dc power. The unfavorable load characteristics are due to the HVDC control methods used, and to the shunt capacitor compensation used to supply reactive power for the converters. For technical and economic reasons, the problems are most pronounced with long-distance transmission. (Also, back-to-back links are usually far removed from major load centers.)

For bulk power system voltage stability, inverters associated with long-distance dc transmission present the most difficulty. AC transmission may parallel the dc transmission. Figure 8-3 shows, schematically, the

Fig. 8-1. 500-kV HVDC converter valves. *Bonneville Power Administration.*

Fig. 8-2. HVDC ac filters.*Bonneville Power Administration.*

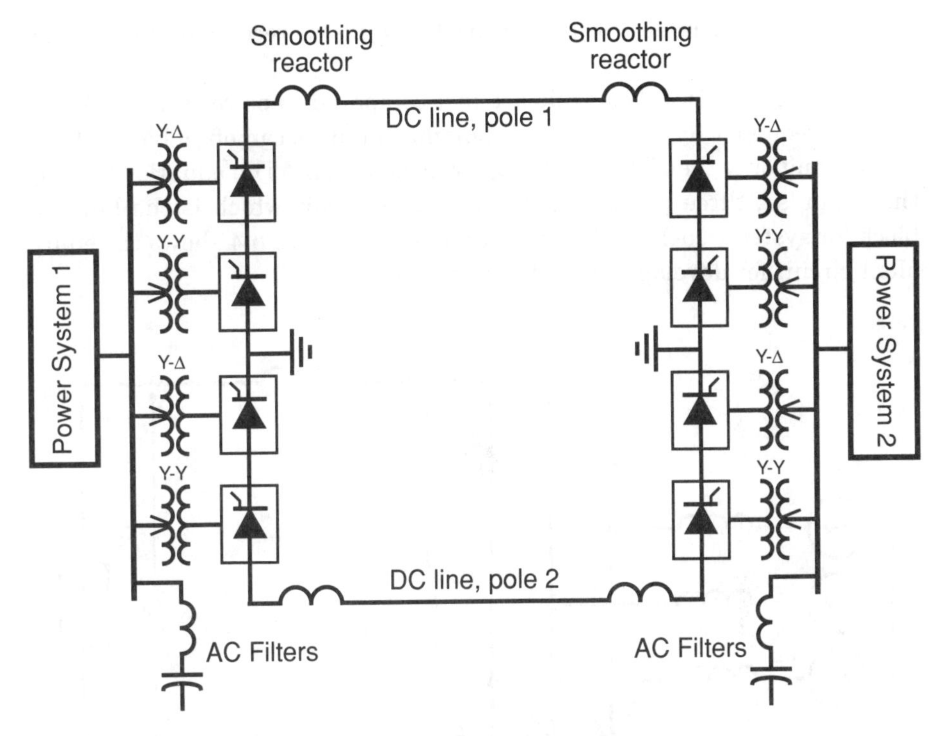

Fig. 8-3. Simplified schematic of conventional two-terminal, twelve-pulse, bipolar, long distance HVDC transmission system. There may be parallel ac transmission.

basic two-terminal, twelve-pulse, bipolar system. The ac filters are capacitive at fundamental frequency.

HVDC-related voltage control (voltage stability and fundamental frequency temporary overvoltages) may be studied using a transient stability program. In some cases an electromagnetic transients program or analog simulator should be used. The effects occur in a fraction of a second or at most a few seconds following a disturbance. Transient stability (including transient damping) are often interrelated with voltage stability.

We will see that voltage instability caused by the HVDC characteristics are similar in several ways to other forms of voltage instability.

8.1 Basic Equations for HVDC

A detailed development of HVDC characteristics is outside our scope. Rather, we will provide the equations and concepts necessary for understanding of basic HVDC operation. Some familiarity with converter theory and HVDC transmission is assumed. We limit our discussion to two terminal, long distance lines. For further background, we recommend references 1–5.

Because the HVDC controls are fast compared to power system fundamental frequency dynamics, quasi-dynamic analysis can often be employed considering only the HVDC link algebraic equations. The equations are for the six-pulse, three phase, full wave bridge circuit which is the building block for systems such as shown on Figure 8-3. Figure 8-4 shows the equivalent circuit for development of the equations [1].

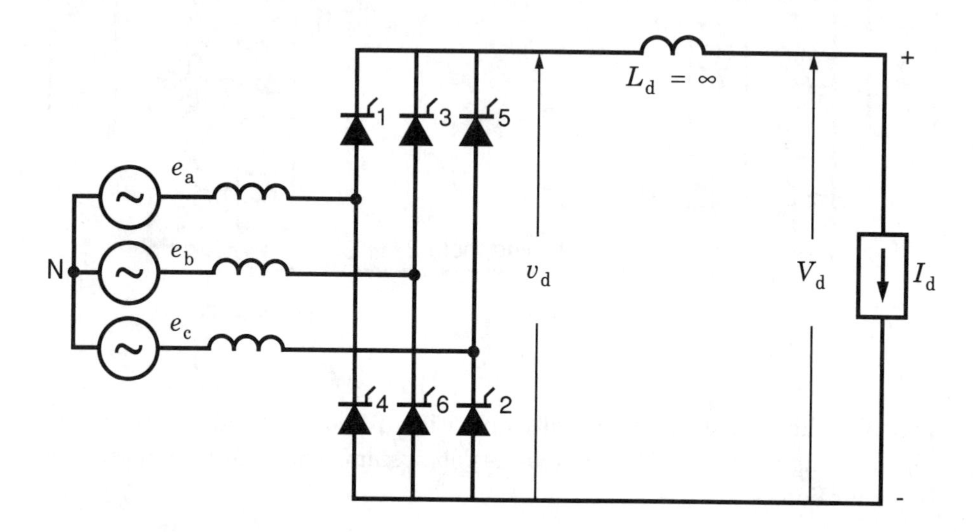

Fig. 8-4. Equivalent circuit for three-phase bridge converter with large dc-side inductor (current source or current "stiff" converter). Numbers indicate valve firing sequence.

We first list the equations (slightly simplified) and then describe their meaning.

Equations for HVDC links:

$$V_{dor} = 1.35TV_{acr}, \; V_{doi} = 1.35TV_{aci}, \; 1.35 = \frac{3\sqrt{2}}{\pi} \tag{8.1}$$

$$V_{dr} = V_{dor}\cos\alpha_r - R_{cr}I_d, \; R_c = \frac{3X_c}{\pi} \tag{8.2}$$

$$V_{di} = V_{doi}\cos\gamma_i - R_{ci}I_d = V_{dr} - R_L I_d \tag{8.3}$$

$$I_d = \frac{V_{dor}\cos\alpha - V_{doi}\cos\gamma}{R_{cr} + R_L - R_{ci}} \tag{8.4}$$

$$I_d = \frac{V_{do}(\cos\alpha + \cos\gamma)}{2R_c} \quad \text{or} \quad \cos\alpha = \frac{2R_c I_d}{V_{do}} - \cos\gamma \tag{8.5}$$

$$P_{dr} = V_{dr}I_d = V_{di}I_d + I_d^2 R_L,\ P_{di} = V_{di}I_d \tag{8.6}$$

$$\cos\phi_r \cong \frac{V_{dr}}{V_{dor}} = \frac{\cos\alpha_r + \cos(\alpha_r + u)}{2} = \cos\alpha_r - \frac{R_{cr}I_d}{V_{dor}} \tag{8.7}$$

$$\cos\phi_i \cong \frac{V_{di}}{V_{doi}} = \frac{\cos\gamma_i + \cos(\gamma_i + u)}{2} = \cos\gamma_i - \frac{R_{ci}I_d}{V_{doi}} \tag{8.8}$$

$$Q_{dr} = P_{dr}\tan\phi_r,\ Q_{di} = P_{di}\tan\phi_i \tag{8.9}$$

Ideal no-load average dc voltage, Equation 8.1. The equation provides the dc voltage for a three-phase bridge rectifier or inverter based on the ac side line-to-line voltage, V_{ac}, and the converter transformer turns ratio T. (The maximum instantaneous voltage is $\sqrt{2}V_{ac}$ and the minimum instantaneous voltage is $\sqrt{2}V_{ac}\cos 30°$. The average dc voltage is obtained by integration over a 60° interval.) Converter transformer under-load tap changing is quite important for HVDC links, and typical time delays are relatively short— typically 5–15 seconds.

The ac side voltage is called the commutating voltage and is usually the converter terminal ac bus voltage. This is because harmonic filters are applied at the converter ac bus to achieve a nearly pure fundamental frequency voltage waveform.

On-load dc voltages, Equations 8.2, 8.3, and 8.4. These equations provide the converter terminal average voltages as affected by firing angle control and by direct current. The ideal no-load voltage is reduced by firing angle, α, control. The inverter equation is written in terms of extinction angle, γ, where $\alpha + \gamma + u = 180°$. A particular valve (thyristor group) has positive bias for 180° during which time the valve thyristors are gated or turned on. The firing (delay) angle is measured from the instant of positive bias. The extinction angle is measured backward in time from the 180° instant. The angular measure of time for transfer (commutation) is the overlap angle, u. The overlap angle is fifteen to twenty degrees at full load. Rectifier firing angle is typically fifteen degrees and inverter extinction angle is typically eighteen degrees. Since commutation must be complete by the 180° instant, the extinction angle is also termed the commutating

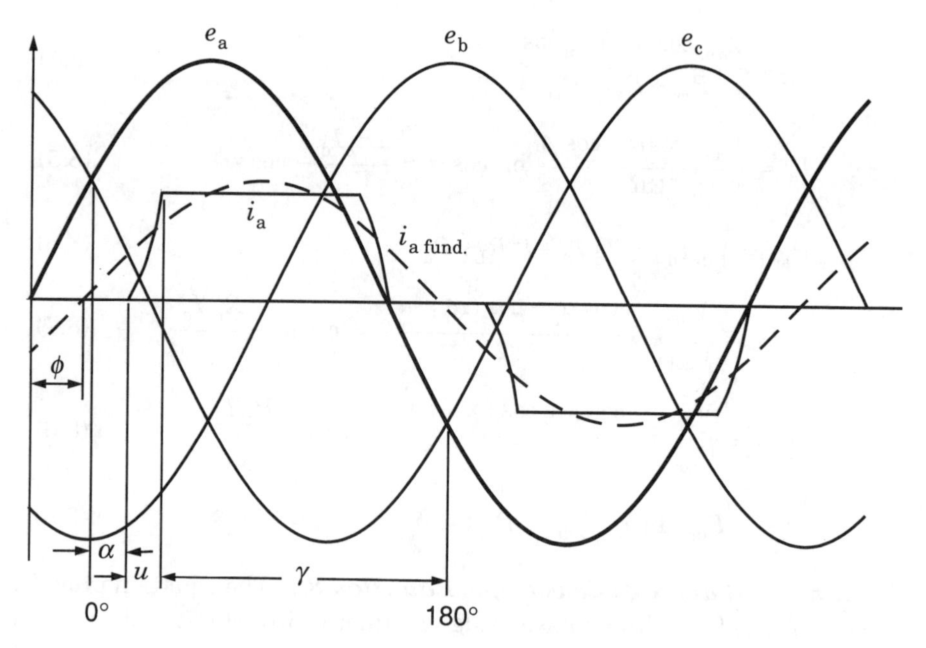

Fig. 8-5. Angle relationships for rectifier conditions. At 0°, phase a becomes positive relative to phase c and commutation from valve 5 to valve 1 starts at the firing angle α. The angular measure of commutation time is u, during which both valves are conducting. At 180° phase a becomes negative relative to phase c. Also shown is the lagging power factor angle, ϕ, between phase a voltage and the fundamental component of phase a current.

margin angle. Figure 8-5 shows the angle relationships for rectifier conditions (small α, large γ).

Because of inductance in the source circuit, commutation cannot occur instantaneously. The nonzero commutation time reduces the dc voltage by the product of "commutating resistance," R_c, and direct current. The commutating resistance is a fictitious resistance that does not consume power. It is proportional to the leakage reactance of the converter transformer, X_c. The direct current, I_d, is constant during steady operation because of the large dc smoothing reactors used; the converters are current sources.

The right-hand-side term of Equation 8.3 reflects the IR drop of the dc transmission line. Equations 8.2 and 8.3 can be solved to obtain Equation 8.4. The negative resistance characteristic for inverters is significant and destabilizing.

Relationship between firing and extinction angles, Equation 8.5. The equation is from converter theory and involves integration over the commutation time [1, page 83]. The factor $2R_c$ results because the circuit involves the commutating reactances of two phases (on Figure 8-2, the circuit for transfer from valve 1 to valve 3 involves the reactances of phases a and b).

Equation 8.5 is important for inverter control. We normally want to control extinction angle, gamma. To avoid "commutation failures," gamma must be greater or equal to a minimum value, γ_{min}. We cannot, however, control gamma directly. Only the firing angle can be controlled. Explicitly or implicitly, the firing angle is determined from the equation.

DC power, Equation 8.6. Power is simply the product of dc voltage and direct current. For long distance transmission, voltage is held at the maximum value and the current is varied for different power schedules. This minimizes losses.

Converter power factor, converter reactive consumption, Equations 8.7, 8.8, and 8.9. Finally we have equations directly relating to voltage stability. We will be examining these equations in detail.

8.2 HVDC Operation

Closed-loop controls regulate rectifier and inverter firing angles to indirectly control variables such as direct current, extinction angle, dc voltage, and dc power. Converter transformer tap changing operates in a slower time frame to optimize operation. Following disturbances, fast intervention by emergency controls may improve power system performance.

Normal operation. For normal operation of long distance, two-terminal HVDC links, the following strategies prevail:

- The rectifier controls dc current through firing angle control.
- The inverter sets the dc voltage through constant (minimum) extinction angle control and ultimately through firing angle control.
- The rectifier tap changer operates to allow desired rectifier firing angle.
- The inverter tap changer operates to control rectifier (sending end) voltage to maximum.
- DC power is controlled at the rectifier as a current regulator outer loop. The ordered current is the scheduled power divided by the dc voltage. The power control may be slow compared to the basic current control.

Disturbance conditions. During emergencies such as ac system faults and severe generator rotor angle swings (which cause voltage swings at the commutating buses), several backup controls are important.

Inverters also have current controllers that operate during low rectifier voltage or high inverter voltage. The inverter current control intervenes after current has dropped about 0.1 per unit and prevents current collapse. The amount the current order is reduced at the inverter is termed current margin.

More important for voltage stability are controls which reduce the direct current during low voltage. An example is a "voltage dependent current order limiter" (VDCOL). By reducing current, voltages are supported as can be seen from the power factor equations. Since dc power is the product of voltage and current, attempting to maintain or increase current during low voltage, is not productive. In extreme situations, the settings and dynamics of the VDCOL can be critical for voltage stability. A good example is the Itaipu HVDC link in Brazil where the inverter station is near the Saõ Paulo load area. Reference 4 describes experience. For the Itaipu link current reduction begins at 0.93 per unit inverter voltage [6]. Voltage stability is a major concern at the Itaipu project.

The VDCOL will help voltage (and power) recovery following faults. In the context of voltage stability following clearing of short circuits, fast voltage recovery helps reacceleration of nearby induction motors.

Several other methods may be used for current or power reduction during low voltage. Detection of low voltage may cause a switchover from power to current control, or a freezing of the voltage measurement used for power control. For the Nelson River HVDC system in Canada, detection of low voltage at the inverter 230-kV ac bus (below 0.95 per unit for 250 ms) results in an HVDC power reduction; this avoids need for additional reactive power support and helps the Winnipeg area voltage stability [7]. The Nelson River power controllers also have a dc voltage hold for ac voltage dips. This puts the system in quasi constant current control while retaining power modulation functions [6,7]. Voltage instability is an important concern at Nelson River, and the dc voltage hold occurs for very small ac voltage dips at the inverters.

A wide variety of special controls are employed to improve power system dynamic performance [6]. These controls can compensate for the unfavorable characteristics of basic dc links. Unlike ac transmission, dc links do not inherently provide synchronizing power in response to a disturbance. As a frequency insensitive load, dc links produce negative damping torques at generators. As we will discuss, dc links with typical controls may contribute to voltage collapse.

The special controls include fast power changes, and modulation of power, current, voltage, or extinction angle. Sometimes there is a conflict between improving synchronous stability and voltage stability.

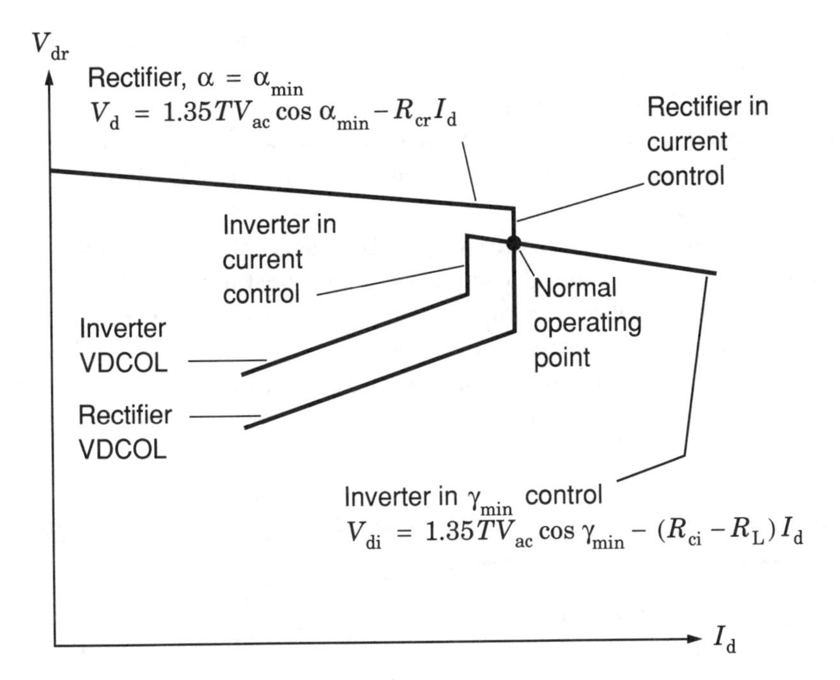

Fig. 8-6. HVDC link steady state characteristic.

Steady-state operating characteristics diagram. We can visualize HVDC operation by static dc voltage–direct current diagrams. Figure 8-6 shows such a diagram, including VDCOL. The voltage is referred to a common point, the rectifier. R_{ci} is usually greater than R_L, resulting in a negative resistance characteristic for the inverter. Current control is shown as a vertical line representing a high gain control system. In firing angle or extinction angle control (shown as α_{min} and γ_{min} control), the slope reflects the commutating resistance and the ac voltage sag as direct current is increased.

Inverter control. The basic inverter control is constant extinction or commutating margin angle control at a minimum value, γ_{min}. The extinction angle is selected to insure sufficient time for commutation during most conditions. Equation 8.8 shows that increasing the extinction angle will decrease power factor. The higher reactive current increases equipment cost and equipment losses. Additional reactive power supply equipment is required.

Minimum extinction angle design and operation results in poor performance, however, especially for weak systems. The inverter characteristics usually modified to improve performance. Figure 8-7a shows the feature which is termed "current error control," "current compounding," or "beta control" ($\beta = 180° - \alpha$). This method helps control system stability since the intersection of the rectifier and inverter characteristics is better defined during disturbance conditions. The positive resistance characteristic also improves power system dynamic performance. Normal operation is still at minimum gamma.

Further improvement requires operation several degrees above the minimum commutating margin angle as shown on Figure 8-7b and 8-7c. The several degrees allows regulating margin. With the inverter in constant dc voltage control, constant power results with the rectifier in constant current control. Complicated multiterminal systems such as the Quebec–New England link use dc voltage control.

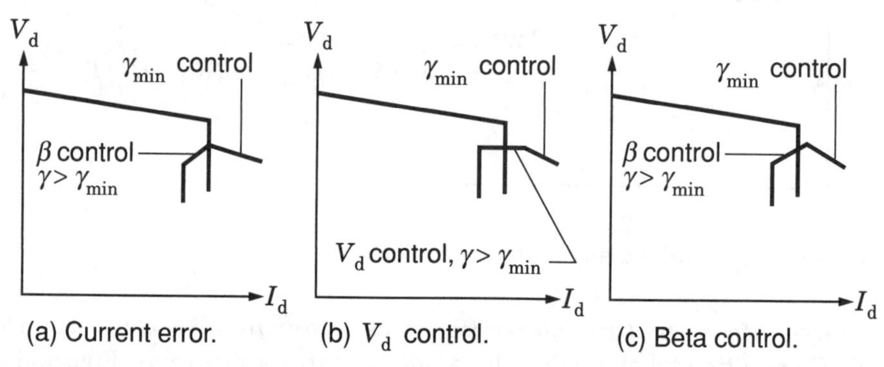

Fig. 8-7. Inverter control methods. Method (a) is commonly used.

Finally, ac voltage can be regulated by gamma modulation. This can be visualized as shifting the inverter static characteristic up and down. If normal operation is at minimum extinction angle, the modulation is one-sided and response is limited to overvoltages excursions.

Several back-to-back links employ ac voltage regulation by gamma control and switching of capacitors, filters, and reactors [10]. The Gezhouba to Shanghai long distance link in China also has this capability [11]. The dc converter valves, valve cooling systems, ac filters, and converter transformers are designed to operate at relatively high firing and extinction angles on a continuous basis.

Example 8-1. For dc voltage control, calculate the dc voltage boost available by operating at an extinction angle of 22°. The minimum extinction angle is 17°.

Solution: Equations 8.1 and 8.3 can be combined to give:

$$V_{di} = 1.35TV_{ac}\cos\gamma - R_{ci}I_d.$$

We are concerned with dips in ac voltage. Since cos 17° = 0.956 and cos 22° = 0.927, a dip of 2.9% can be compensated. There is much more capability to compensate for overvoltages by increasing the extinction angle.

Example 8-2. An HVDC link is rated and operated at 1000 MW. The overlap angle at full load is 20°. Calculate the reactive power consumption for operation at a minimum extinction angle of 17° and for operation at 22°.

Solution: Equations 8.8 and 8.9 apply. For $\gamma = 17°$, $\cos\phi = 0.8775$ giving a reactive consumption of 547 MVAr. For $\gamma = 22°$, $\cos\phi = 0.8352$ and the reactive consumption is 659 MVAr. Operation at 22° with ac voltage control (gamma modulation) is equivalent to a static var compensator with 112 MVAr capacitive range and with a large inductive range. For large voltage dips, the gamma modulation could be combined with mechanically switched capacitors.

8.3 Voltage Collapse

We now develop a scenario for voltage collapse involving an inverter in constant extinction angle control.

1. A large disturbance occurs near the inverter during heavy load. AC interconnections may be heavily loaded. The power system is highly stressed.

2. The disturbance causes voltage sag, and voltage recovery may be slow. Subsequent voltage swings result from electromechanical oscillation between generators.

3. Reactive power output of capacitor banks and filters are reduced during low voltage.

4. In some cases, there are controls that increase extinction angle for ac voltage drop. This is to maintain constant volt-second commutating margin.

5. The inverter power factor will be worse because of the low ac voltage drop and the possible gamma increase. This is apparent from Equation 8.8 which we rewrite:

$$\cos\phi_i = \cos\gamma_i - \frac{R_{ci}I_d}{1.35TV_{ac}}.$$

6. This results in higher reactive current demand. If power control is fast compared to the voltage swing or attempted voltage recovery, direct current will be increased—making the power factor still lower. The reactive power consumption will increase according to $Q = P\tan\phi$ (Equation

8.9). Attempting a fast dc power increase to improve transient rotor angle stability will further increase reactive power demand and may be counterproductive.

7. The combined effect of the reduced capacitor bank and filter outputs and the increased inverter reactive current demand will cause the ac voltage to drop more, possibly leading to collapse.

8. Converter transformer tap changing may occur within ten seconds. This may put more burden on the ac system.

9. Reduction in dc current and power by VDCOL or other controls may stabilize the system voltage. Reduction in dc power, however, may aggravate transient rotor angle stability, leading to more severe voltage swings.

Note that even if the inverter was initially in constant dc voltage control, the minimum extinction angle mode would result because of declining ac voltage and increasing direct current.

8.4 Voltage Stability Concepts Based on Short Circuit Ratio

Although we have just presented the scenario for voltage collapse involving HVDC links, additional insight on voltage stability can be gained by exploring concepts having parallels in purely ac systems.

Short Circuit Ratio (SCR) and Effective Short Circuit Ratio (ESCR). We defined these terms in Chapter 1, Section 5. For the application of HVDC links, they are commonly used as simple methods to relate dc power transfer to the strength of the source ac system. Recall that the short circuit ratio is the system short circuit capacity divided by the dc power. A low short circuit ratio requires special techniques to achieve satisfactory system dynamic performance.

Most of the reactive power required by a converter is supplied at the terminal. Effective short circuit ratio (ESCR) includes the effect of HVDC terminal reactive power equipment on short circuit capacity. The equipment may include synchronous condensers, shunt capacitor banks, and harmonic filters (which are capacitive at fundamental frequency). Compared to the basic short circuit ratio from the network, the effective short circuit ratio is increased by synchronous condensers, and reduced by capacitor banks and filters. Synchronous condensers, although used infrequently, obviously strengthen the network. Capacitors are voltage destabilizing. These facts are reflected in the ESCR.

Example 8-3. An inverter supplying one per unit power to a load area consumes 0.56 per unit reactive power. The system short circuit capacity is

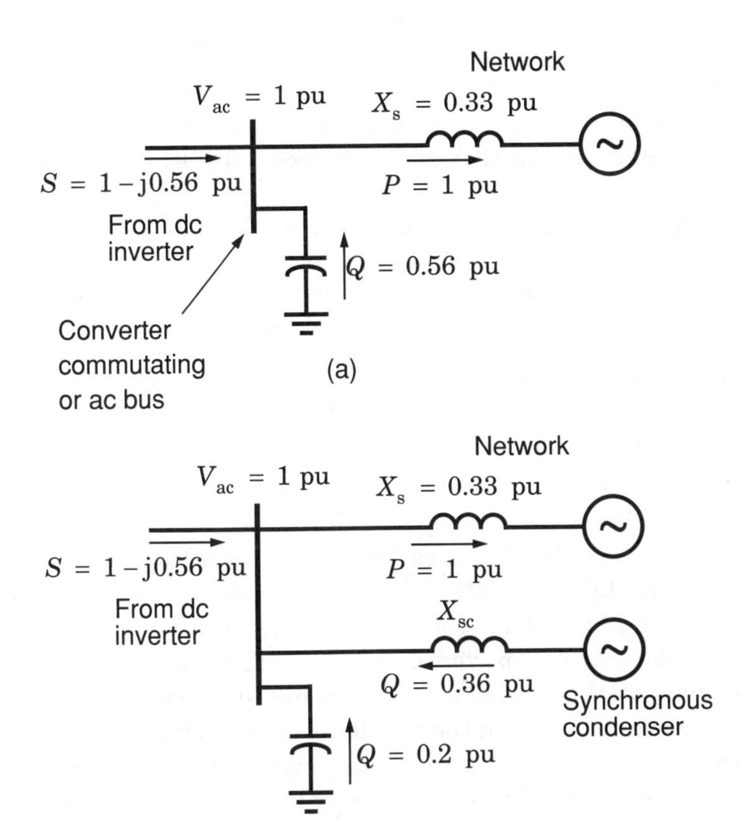

Fig. 8-8. Circuits for example 8-3.

three per unit; therefore the basic short circuit ratio is 3.0. Figure 8-8 shows two methods to provide the required reactive power. For the synchronous condenser case, the condenser and stepup transformer impedance totals 0.4 per unit on the condenser base (condenser base is 0.36 per unit of the dc base power). Calculate the effective short circuit ratio for each method.

Solution for case a (shunt capacitor banks): The capacitive reactance is 1/0.56 or 1.786. The thévenin impedance is:

$$X_{th} = \frac{j0.33(-j1.786)}{-j(1.786-0.33)} = j0.4048.$$

The corresponding short circuit power is $V^2/X_{th} = 1/0.4048 = 2.47$. The ESCR ratio is thus 2.47 compared to the basic short circuit ratio of 3.

Solution for case b (synchronous condenser): The condenser reactance on the system (dc power) base is 0.4/0.36 = 1.111. The parallel combination of the condenser and the source reactance is 0.2544. The shunt capacitor reactance is 1/0.2 = 5. The thévenin reactance is 0.2681 and the ESCR ratio is 3.73.

A simpler method of calculating ESCR is:

$$\text{ESCR} = \frac{\text{system sc power} + \text{condenser sc power} - \text{capacitor MVAr}}{\text{dc power}}$$

Short circuit ratios can be classified as follows:

High	SCR > 5
Moderate	SCR = 3–5
Low	SCR = 2–3
Very Low	SCR < 2

In recent years, HVDC links have been installed with short circuit ratios as low as 1.6. The McNeill back-to-back station in Alberta, Canada has a minimum ESCR of less than 1.0 [12]. Special techniques are required to achieve satisfactory performance. An IEEE Committee report [8] describes five installations all of which are back-to-back links. Special techniques are easier to apply on the simpler back-to-back links.

In some cases, we are interested in power system disturbances affecting several HVDC links in close proximity to each other. For example, two high capacity long-distance links terminate in the Los Angeles load area (Pacific HVDC Intertie and Intermountain Power Project). In this case, we can consider a "combined SCR" where the power system has to support the unfavorable load characteristics of several links.

Example 8-4. For voltage stability in the Los Angeles load area, calculate the combined SCR if the minimum short circuit capacity is 14,000 MVA.

Solution: The rating of the Pacific HVDC Intertie (Celilo to Sylmar) is 3100 MW and the rating of the Intermountain Power Project (Intermountain to Adelanto) is 1920 MW. Since the Los Angeles terminals (Sylmar and Adelanto) are normally inverters, the received power is less than the ratings because of dc transmission line losses—this effect will be neglected. The combined SCR is then 14,000/(3100 + 1920) or 2.8. Since only capacitor banks and filters are used for reactive supply, the effective short circuit ratio is lower. This measure is highly approximate and depends, among other factors, on how closely the two links are coupled for a particular disturbance.

Several years ago, a third dc link into the Los Angeles load area was considered (Phoenix to Mead to Adelanto 2000 MW link). This would have reduced the combined short circuit ratio to about 2.0.

Maximum power and power instability. In Chapter 2, for ac systems, we discussed the idea of maximum power. For a basic system, power is maximum when the magnitude of load impedance equals the magnitude of source impedance. Adding more load results in less power.

HVDC links have analogous characteristics. As direct current is increased, dc power will increase. At some point, however, further increase in direct current will cause lower dc power. The reason is simply that the increased reactive power consumption sags ac voltage, and thereby dc voltage, so that the dc voltage is falling more than the direct current increase. We could show this on a $P_d - V_{ac}$ nose curve and on a P_d versus I_d curve. Compared to *P–V* curves for a simple ac system, curves for an inverter in constant extinction angle control feeding a source system will cut across the constant power factor lines of Figure 2-8; the power factor becomes more lagging as dc power is increased [13,14].

On the unstable side, continued current increase by power control would cause voltage collapse. Usually, however, other controls would act to limit or reduce current.

Figure 8-9 shows the sensitivity of dc power to changes in direct current as a function of inverter terminal short circuit ratio [15]. The thévenin

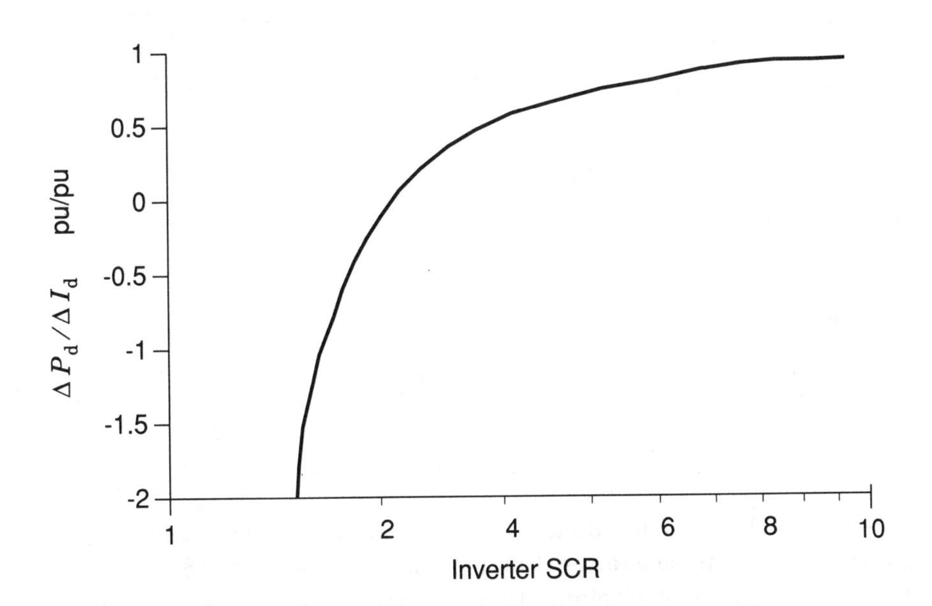

Fig. 8-9. Sensitivity of dc power to changes in direct current. Inverter in constant extinction angle control at $\gamma = 18°$. Commutation reactance is 13% and source impedance angle is 90° [15]. © 1986 IEEE.

source voltage is one per unit and one per unit voltage is obtained at the inverter terminal by application of shunt capacitor banks. The inverter is in constant extinction angle control. For a high short circuit ratio, the per unit power increase is nearly equal to the per unit current increase. For a short circuit ratio less than about 2.2, an increase in direct current results in lower dc power. DC power control at low short circuit ratios results in *power* instability.

Tap changer instability. In conjunction with Figure 2-14, we discussed tap changer instability. At the nose of a P–V curve, tap changing to reflect additional load conductance to the primary system results in reduced power, meaning the load-side voltage has decreased.

Similar phenomena exists with HVDC links supported by weak ac systems. For the same conditions as for Figure 8-9, Figure 8-10 shows sensitivity of dc voltage to a 1% change in converter transformer tap ratio for varying short circuit ratios [15]. For high short circuit ratios, nearly a 1% change in voltage is obtained. For short circuit ratios less than about 1.9, the voltage change is negative.

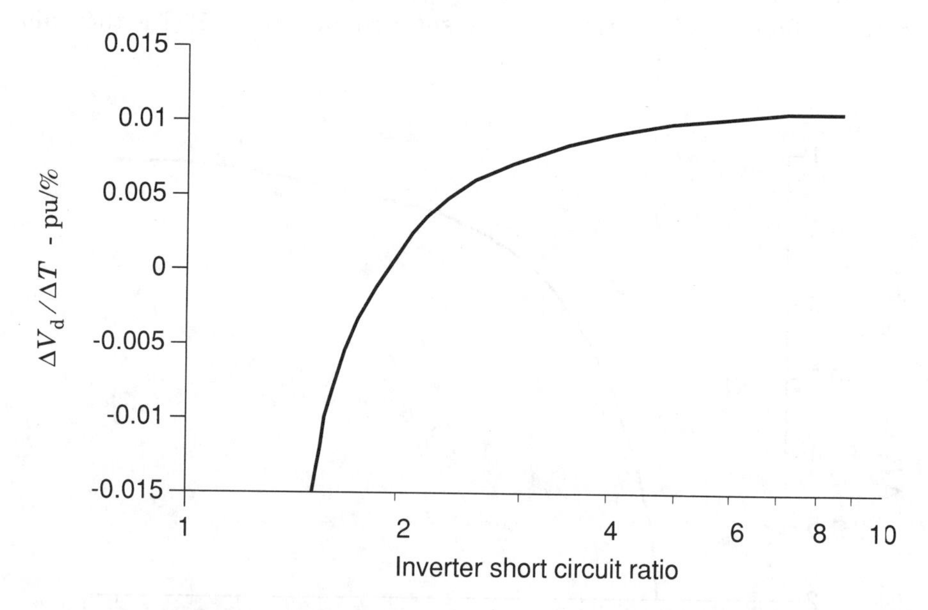

Fig. 8-10. Sensitivity of dc voltage to a 1% change in converter transformer tap ratio. Inverter in constant extinction angle control at γ=18° and rectifier is in current control. Commutation reactance is 13% and source impedance angle is 90° [15]. © 1986 IEEE.

Voltage Stability Factor (VSF). Short circuit ratio measures can also be defined with regard to reactive power consumed by an HVDC converter. This is appealing since it is the variation of reactive power that causes the major change in system voltage. This is captured in a less simplistic manner by "voltage stability factor."

Short circuit capacity related measures are simple. For some purposes they are useful—for other purposes they are simplistic. They do not account for the fast-acting controls of HVDC links and static var compensators.

Short circuit ratios also do not clearly indicate problems with voltage control during high ac system power transfers. We can see this effect on *P–V* or *Q–V* curves—as we get closer to the nose of the curve, the voltage change for a power increment increases.

Although dynamic study is normally used for detailed analysis, Hammad and Kühn developed a power flow program based "Voltage Stability Factor" (VSF) measure of AC/DC voltage control and stability [16]. The Voltage Stability Factor is defined as:

$$\mathrm{VSF} = \left.\frac{\Delta V_{\mathrm{ac}}}{\Delta q}\right|_{P_{\mathrm{d}}} = \left.\frac{1}{\dfrac{\Delta Q_{\mathrm{a}}}{\Delta V_{\mathrm{ac}}} + \dfrac{\Delta Q_{\mathrm{d}}}{\Delta V_{\mathrm{ac}}} - \dfrac{\Delta Q_{\mathrm{c}}}{\Delta V_{\mathrm{ac}}} - \dfrac{\Delta Q_{\mathrm{s}}}{\Delta V_{\mathrm{ac}}}}\right|_{P_{\mathrm{d}}} \tag{8.10}$$

where $\Delta q = \Delta Q_{\mathrm{a}} + \Delta Q_{\mathrm{d}} - \Delta Q_{\mathrm{c}} - \Delta Q_{\mathrm{s}}$ and ΔV_{ac} is a variation of voltage at the converter commutating bus (converter ac bus). ΔQ_{a} is the incremental reactive power from the ac network, ΔQ_{d} is the incremental converter reactive consumption, ΔQ_{c} is the incremental reactive power from capacitor banks and filters, and ΔQ_{s} is the incremental reactive power from an optional static var compensator or synchronous condenser. See Figure 8-11.

Following every power flow solution, the VSF can be computed using sensitivity computation with the linearized system. The equations are modified using appropriate models such as voltage dependent load models and voltage behind transient reactance generator models. The denominator terms of the right-hand-side term of Equation 8-10 are linearized representation of the controlled equipment; for example, the $\Delta Q_{\mathrm{s}}/\Delta V_{\mathrm{ac}}$ term for a static var compensator is the inverse of the slope characteristic. The method of inverter control has a major effect on the VSF.

Figure 8-12 shows VSF results for an inverter with short circuit ratio of 1.8. Three inverter control modes are shown: constant extinction angle, constant beta (firing angle) control, and ac voltage control with switched shunt capacitor banks. The small-disturbance voltage stability is greatly improved by the latter method. Large disturbance performance for low

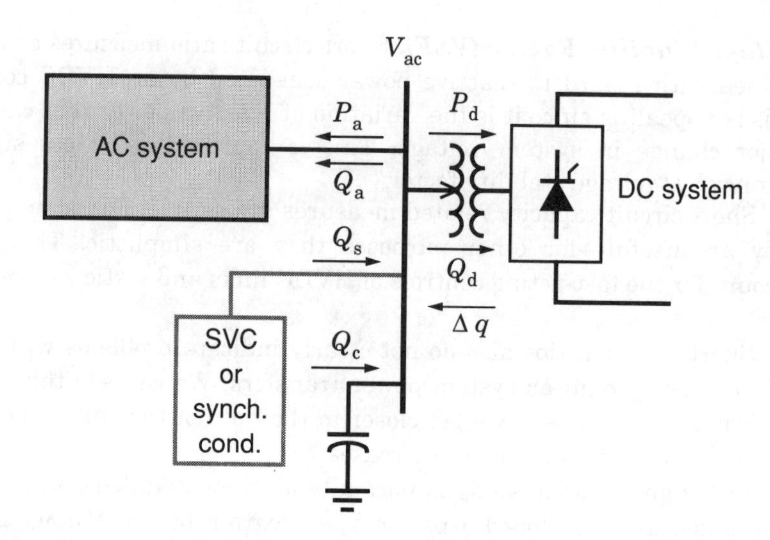

Fig. 8-11. System components to determine Voltage Stability Factor [16]. © 1986 IEEE.

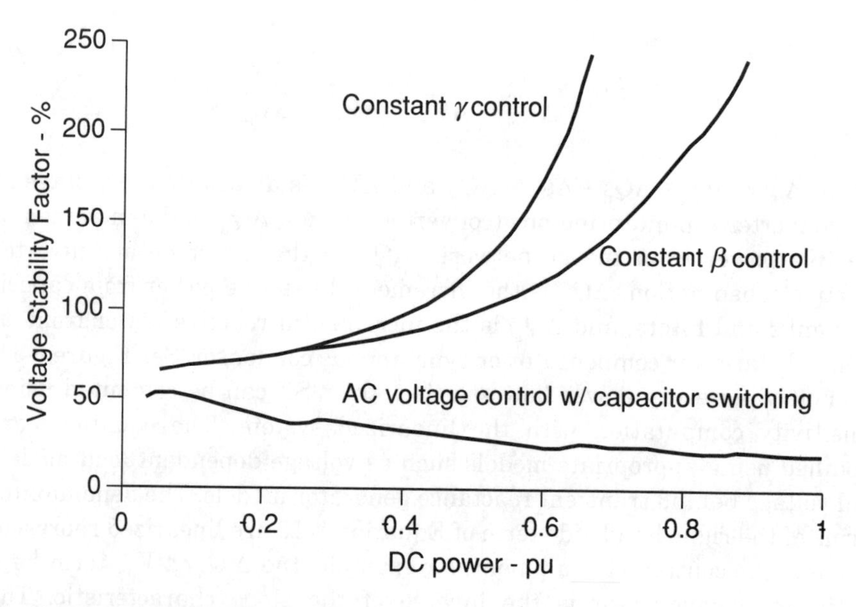

Fig. 8-12. Voltage Stability Factor for three types of inverter control [16]. SCR = 1.8. © 1986 IEEE.

voltage would depend on the available mechanically switched capacitor banks.

8.5 Power System Dynamic Performance

Analysis of realistic systems usually involves simulation using a transient stability program. Detailed models for the HVDC link and for motor loads are required. Often both synchronous and voltage stability must be considered, and appropriate trade-offs made between the two types of stability. The interaction between synchronous stability (transient stability including transient damping) and voltage stability may be very complex.

Comparison between AC and DC Transmission. In planning a new long distance interconnection, we have to choose between HVDC and ac transmission. Dynamic performance (synchronous and voltage stability) should be one of the considerations in the decision. The ability to apply fast-acting stability enhancing controls to HVDC links is often stated as an important advantage of HVDC. We must examine this "common wisdom" carefully.

With ac transmission, large networks operate in synchronism. Following disturbances, the inherent properties of ac transmission support the stability of interconnected synchronous generators. For example, for outage of a line, parallel lines will virtually instantaneously pick up much of the power that was carried on the lost line. Inertial power swings will follow to provide additional synchronizing and damping torques to generators. Although reactive power losses will increase, and line charging power will decrease, the effects are relatively minor in the transient stability time frame, particularly if load characteristics are not especially onerous.

DC transmission in the usual constant power control mode does not provide additional power without special controls. Robust closed-loop special controls may be difficult to design [17–19]. For disturbances and low voltages, power control increases direct current—significantly increasing the reactive demand of converters. We have learned that the combination of power control, constant extinction angle control at inverters, and shunt capacitors may threaten voltage stability. Switching from power control to current control will help voltage stability at the expense of synchronous stability. Power modulation for synchronizing or damping support will be lost in current control.

For the western North American interconnection, we compared ac and dc transmission [17]. The transient stability simulation comparison involved replacing the 787 km Intermountain Power Project HVDC link with a hypothetical ac link. For the two disturbances studied, system dynamic performance with the ac link dynamic performance was remarkably superior. For the cases with the dc link, the effect of switching from power control to current control during low voltage was investigated—for one disturbance it was beneficial, for the other it was detrimental.

Simulation study with voltage collapse. In Example 8-4, we mentioned that a third HVDC link into the Los Angeles area was considered several years ago. The 2000 MW link was three-terminal—from the Phoenix, Arizona area to the Mead Substation near Las Vegas, Nevada to the Adelanto Substation east of Los Angeles. The Intermountain Power Project inverter is also at Adelanto.

Simulation studies are reported in reference 20. The disturbance studied was a three-phase fault near the Palo Verde Nuclear Plant west of Phoenix. The fault was on a 500-kV ac line from Palo Verde to the Devers Substation east of Los Angeles near Palm Springs (permanent outage of the line). This line is in parallel with the proposed three-terminal HVDC link. The Intermountain Power Project was modeled as operated: rectifier in power control and inverter in constant extinction angle control.

Simulation results showed system dynamic performance benefited more from reduction of converter reactive power demand than from fast dc power restoration and synchronizing support.

For the Palo Verde fault and line outage, system stability was marginal prior to adding the additional dc link. Adding the new dc link caused a fast voltage collapse (before significant rotor angle swings).

With high level modulation of dc power based on Adelanto ac voltage, transient instability and voltage collapse occurred on the first swing.

The power modulation reduced dc power in phase with, and in proportion to, the ac voltage drop at the Adelanto inverter. Telemetry was required to the main rectifier at Phoenix. The effect was to release reactive power at both the Phoenix area rectifier and the Adelanto inverter. The power modulation, while preventing fast voltage collapse, had a de-synchronizing effect.

A stable case required, in addition to the power modulation, a 400 MVAr static var compensator and gamma modulation at the Adelanto inverter.

Figure 8-13 shows ac voltage at Adelanto for three cases: pre-dc link, power modulation alone, and power modulation plus SVC and gamma modulation. For the first swing of the stable case, the power modulation reduced the Adelanto power and current by about 70%. This reduced the Adelanto inverter reactive power demand by about 800 MVAr.

References

1. E. W. Kimbark, *Direct Current Transmission*, Wiley-Interscience, 1971.
2. Ebasco Services Incorporated, *Methodology for Integration of HVDC Links in Large AC Systems—Phase 1: Reference Manual*, EPRI Final Report EL-3004, March 1983.

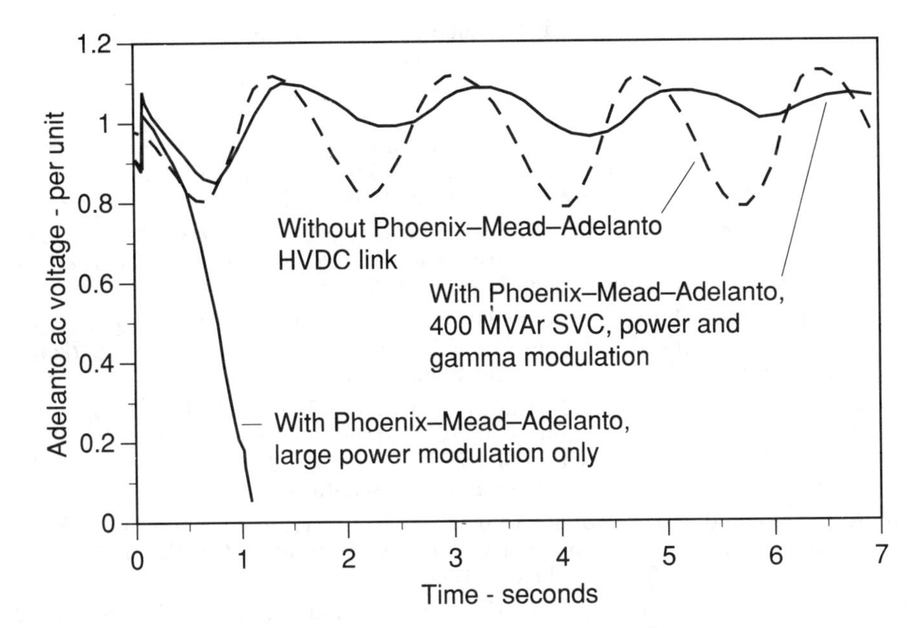

Fig. 8-13. Adelanto ac voltage for outage three-phase fault at Palo Verde and outage of the Palo Verde–Devers 500-kV line. 1992 light load summer conditions [20]. © 1988 IEEE.

3. Institut de Recherche d'Hydro-Québec, *Methodology for the Integration of HVDC Links in Large AC Systems—Phase 2: Advanced Concepts*, EPRI Final Report EL-4365, April 1987.
4. P. Kundur, *Power System Stability and Control*, McGraw-Hill, 1993.
5. CIGRÉ Working Group14-07 and IEEE Working Group15.05.05, *Guide for Planning DC Links Terminating at AC Systems Locations Having Low Short--Circuit Capacities, Part I: AC/DC Interaction Phenomena*, CIGRÉ, June 1992.
6. IEEE Committee Report, "HVDC Controls for System Dynamic Performance," *IEEE Transactions on Power Systems*, Vol. 6, No. 2, pp. 743–752, May 1991.
7. R. W. Haywood and J. Chand, "Responses of the Nelson River HVDC System to Disturbances on the Receiving End AC Network," *CIGRÉ, Proceedings of 31st Session*, Vol. I, paper 14-04, 1984.
8. M. M. Rashwan, G. B. Mazur, M. A. Weekes, and D. P. Brandt, "AC/DC System Power/Voltage Stability Enhancement Using Modified Power Control," *Proceedings of Second HVDC System Operating Conference*, pp. 35–39, Winnipeg, Canada, 18–21 September 1989.
9. J. Chand and D. Tang, "Operational Experience of the Nelson River HVDC System," *Proceedings of Second HVDC System Operating Conference*, pp. 21–31, Winnipeg, Canada, 18–21 September 1989.
10. IEEE Committee Report, "DC Transmission Terminating at Low Short Circuit Ratio Locations," *IEEE Transactions on Power Delivery*, Vol. PWRD-1, No. 3, pp. 308–318, July 1986.

11. A. Hammad, H. Koelsch, and P. Daehler, "Active and Reactive Power Controls for the Gezhouba–Shanghai HVDC Transmission Scheme," *Fifth International Conference on AC and DC Power Transmission*, IEE Conference Publication No. 345, pp. 279–284, September 1991.
12. R. P. Burgess, J. D. Ainsworth, H. L. Thanawala, M. Jain, and R. S. Burton, "Voltage/VAr Control at McNeill Back-to-Back HVDC Convertor Station," *CIGRÉ, Papers of the 1990 Session*, paper 14-104, 1990.
13. A. Gavrilovic, et al. (CIGRÉ/IEEE committee paper), "Interaction Between DC and AC Systems," *CIGRÉ Symposium on AC/DC Transmission Interactions and Comparisons*, paper 200-20, Boston, 28-30 September 1987.
14. L. A. S. Pilotto, M. Szechtman, and A. Hammad, "Transient AC Voltage Related Phenomena for HVDC Schemes Connected to Weak AC Systems," *IEEE Transactions on Power Delivery*, Vol. 7, No. 3, pp. 1396–1404, July 1992.
15. R. J. Piwko, F. Nozari, R. L. Hauth, and C. W. Flairty, "Control Systems for Application in HVDC Terminals at AC System Locations Having Low Short Circuit Capacities," *Proceeding of IEEE MONTECH'86 Conference on HVDC Power Transmission*, Montreal, September 29–October 1, 1986.
16. A. E. Hammad and W. Kühn, "A Computation Algorithm for Assessing Voltage Stability at AC/DC Interconnections," *IEEE Transactions on Power Systems*, Vol. 1, No. 1, pp. 209–216, February 1986.
17. IEEE Committee Report, "AC-DC Economics and Alternatives—1987 Panel Session Report," *IEEE Transactions on Power Delivery*, Vol. 5, No. 4, pp. 1956–1976, October 1990.
18. J. F. Hauer, discussion of "AC-DC Economics and Alternatives—1987 Panel Session Report," *IEEE Transactions on Power Delivery*, Vol. 5, No. 4, pp. 1956–1976, October 1990.
19. R. L. Cresap, D. N. Scott, W. A. Mittelstadt, and C. W. Taylor, "Operating Experience with Modulation of the Pacific HVDC Intertie," *IEEE Transactions on Power Apparatus and Systems*, Vol. PAS-98, pp. 1053-1059, July/August 1978.
20. R. L. Lee, D. Zollman, J. F. Tang, J. C. Hsu, J. R. Hunt, R. S. Burton, and D. E. Fletcher, "Enhancement of AC/DC System Performance by Modulation of a Proposed Multiterminal DC system in the Southwestern U. S.," *IEEE Transactions on Power Delivery*, Vol. 3, No. 1, pp. 307–316, January 1988.

9

Power System Planning and Operating Guidelines

No problem can stand the assault of sustained thinking.
Voltaire

In both planning and operation, the main goal in studying voltage stability is to increase power transfer by removing voltage stability limits. If we remove voltage stability as the limit, we work on the new limitation (perhaps transient stability) which may result in voltage stability limits again. Ultimately, thermal limits may be reached.

We not only want to analyze voltage stability, we want to *solve* voltage stability problems.

Voltage stability has many facets. There are also many solutions to voltage stability challenges. The solutions involve all types of equipment—generation, transmission, and distribution. We want to find low cost solutions wherever possible; this often means special controls. Solutions also involve power system operating methods. Solutions described in previous chapters will be summarized in this chapter and some new methods will be described.

Before describing solutions, we will discuss planning and operating criteria. What constitutes acceptable performance?

9.1 Reliability Criteria

In designing and operating a power system, criteria should be developed, and agreed to, by involved parties. The criteria guides us as to what performance is required. Reliability of 100% is not possible because unusual disturbances sometimes occur. Attempting 100% reliability is not desirable because the cost would be very high. A prudent amount of risk must be taken.

Planning (design) and operating criteria may be either deterministic or value-based (probabilistic). For stability, deterministic criteria are most common. Based on experience and judgement, certain "credible" disturbances are selected to test stability. Certain performance is then required.

Disturbance criteria. For transient rotor angle (synchronous) stability, a typical criterion in developed countries is to remain stable, with some margin, for a three-phase fault on a critical line near a major generating station; the fault is cleared in normal relay and circuit breaker time by opening the faulted line. The required margin to instability may be based on voltage dip during swings, or based on stability with interties loaded some percentage above the operating limit. Although three-phase faults on EHV lines near generating stations are rare, the three-phase fault "umbrella" ensures stable operation for a variety of multiple contingency disturbances.

Value-based criteria, or aspects of value-based criteria, will probably be used in the future to optimize design and operation. This requires consideration of statistics on frequency of faults, fault type, fault location; frequency of line and multiple line outages with or without successful reclosing; and frequency of generation outages. Factors such as load levels and relay misoperation probabilities are important. Costs of instability must be estimated. By using much more information specific to the problem at hand, overdesign may be avoided.* Prudent risk is taken. In computing the cost of instability, attention is focused on backup protection to prevent or mitigate wide-spread blackouts (e.g., automatic load shedding, controlled separations, or plans for fast restoration).

Reliability criteria for voltage stability are not as well developed as criteria for rotor angle stability. One concept is to establish voltage stability criteria with similar probabilities as those used for rotor angle stability [2]. For example, the three-phase fault criterion described above has low probability. For longer-term voltage stability, short circuits are not important—only the resulting line outage. For similarly severe low probability disturbances affecting voltage stability, multiple contingencies are required. This could be outage of several lines (especially double-circuit lines), or a line outage with reactive sources or load area generators out-of-service. Other conditions such as load level and generation pattern, and modeling assumptions could be factored into criteria.

---

*Quoting from Pereira and Balu [1]: "...the deterministic criteria may lead to uneconomic designs (usually overdesign) as compared with the "true" optimal design, which takes into account the *probability* of each operating scenario, and the *economic consequences* of underperformance under the scenarios."

Example 9-1. For voltage stability in the Puget Sound area of the Pacific Northwest, the following criteria are being used: For heavy loads (one-in-two years cold weather), stability for a critical double-circuit 500-kV line outage is required. For extra heavy loads (one-in-twenty years very cold weather), stability for a critical single-circuit 500-kV line outage is required. For the two events, the joint probability of load level and disturbance are somewhat similar.

Performance criteria and margin to instability. Performance criteria also are needed. Traditionally, post-disturbance voltage magnitudes requirements have been used. Voltages must not be below a specified value such as 0.95 per unit. This is not satisfactory, since reactive power reserves could be very low. The combination of voltage magnitude and reactive power reserves *at effective locations* provides good planning or operating criteria.

A required margin or security measure from an operating point to instability is also used. (Usually, the margin is to the maximum power transfer point rather than the instability point, with the tacit assumption that operation at higher loads is unacceptable.) Margin allows for inaccuracies in simulation, and for differences between actual and simulated operating conditions. Experience, judgement, and sensitivity studies are required in selecting appropriate margin for a particular power system. For economic reasons, excessive margin requirements must be avoided; overly conservative (pessimistic) models must also be avoided [3]. Various candidate variables for margin requirements are interlinked and relate in some manner to adequate reactive power reserves.

Considering P–V or Q–V curves, margin can be quantified in terms of MW (or MVA) or MVAr distance from the operating point to the critical or maximum power transfer point (Figure 9-1). MW margin can be in terms of a transmission interface or in terms of area load. Usually, area load is assumed to increase with constant power factor. MVAr margin often is computed for a single key bus using the fictitious synchronous condenser method (V–Q rather than Q–V curve). For large systems, however, computing reactive power margin by reactive load increases throughout an area is more realistic.

MW margin could allow for loads exceeding forecasts. Alternatively, MW margin could relate to forced or prior outages of generation in a load area; this means that stability may be maintained for an additional contingency. If a 500 MW margin is used, stability should be maintained for loss of generators smaller than 500 MW.

MVAr margins are attractive since voltage and reactive power are closely linked. The MVAr margin could relate to the size of shunt capacitor banks or static var compensators in the area. If a 150 MVAr shunt capacitor bank is nearby, a margin greater than 150 MVAr ensures that stability is maintained for the additional contingency of loss of nearby reactive power sources. MVAr margin also allows for loads exceeding forecast. During heavy loads, each additional megawatt of load may require two or three megavar of reactive power support.

As discussed in reference 2, the MW or MVAr margin could relate to a fairly high probability situation such as a load-area generator being offline.

The post-disturbance margin must be satisfied by the system design, or by the pre-disturbance system operation.

The *P–V* curve slope ($\Delta V/\Delta P$) or the *V–Q* curve slope ($\Delta V/\Delta Q$) at the operating point may be used as an indicator of voltage security. A large change in voltage for a small change in loading means abnormally high stress, with the possibility of current limiting at some generators.

P–V and *V–Q* curves provide a critical voltage. In order for voltage magnitude monitoring to be meaningful, the critical voltage must be below normal operating voltages. In the past, British Columbia Hydro used performance criteria requiring that the critical voltage must be below 0.95 per unit and that the critical voltage must be 5% below the operating voltage [4, page 152]. Assuming traditional constant power load models, these requirements may be very difficult to meet. The criteria can rule out voltage stability solutions involving shunt capacitor banks or static var compensators that improve MW or MVAr margin but raise critical voltage.

The amount of reactive reserves, particularly fast-acting reserves from generators or static compensators, affects voltage stability. The reactive

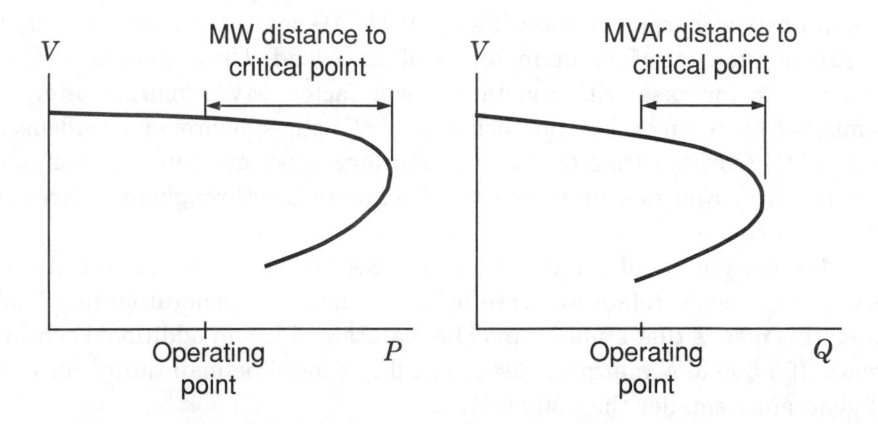

Fig. 9-1. MW and MVAr distance to voltage instability (maximum power or critical point).

reserves must be at effective locations. If meaningful "voltage control areas" can be defined, adequate reactive reserve must be available in the voltage control area in question. Current limiting at any generator or SVC reduces security.

As described in Chapters 6 and Appendix B, sensitivity analysis and steady-state modal analysis [5] provide considerable insight into critical locations for reactive reserves. Appendix B also describes *indices* for voltage stability.

For any selected criteria, the models and methods of analysis are important. For example, a *P–V* curve may be computed for a contingency by starting from a base case, taking the contingency, and then increasing power until the nose is reached; the MW distance to voltage collapse is a measure of system robustness. Alternatively, a new base case with appropriate generation dispatch could be computed at each power level; contingency cases for that power level are then computed. If load build-up is the concern, contingency cases may not be required.

Guidelines. Reliability criteria for voltage stability are evolving and it will be several years before criteria starts to be standardized. The nature of voltage stability problems for different utilities and systems may be such that several approaches will be used. This is true today, as some utilities rely mainly on *P–V* curve performance and others rely mainly on *V–Q* curve performance. Modeling approaches also differ.

Nevertheless, we will make certain recommendations. The recommendations are largely based on voltage stability analysis of previous chapters. See also Appendix C.

1. For large-scale systems, the *V–Q* curve method applied to a single test bus is unrealistic and should not be used for decision making. Although widely used in the past, new computer programs facilitate improved analysis techniques.

2. Disturbance criteria should consider other factors besides the primary line or generator outage. Prior outage of reactive sources, generation, or reduced reactive capability of some generation may be appropriate. The probability of losing more than one line should be evaluated, particularly for double-circuit lines or lines terminating at the same substations. Sympathetic relaying (relay misoperation on a parallel unfaulted line) has always been a problem. One method to provide for prior outages of equipment is an appropriate MW or MVAr margin to the maximum power transfer capability point.

3. Disturbance and performance criteria severity should account for backup control and protection in service. Backup control may include voltage control of shunt capacitors banks normally switched by operators. Cri-

teria can be relaxed if small amounts of undervoltage load shedding stabilizes the system with acceptable voltages.

4. Modeling of loads and generation should be carefully considered. Constant power modeling of loads may be too conservative, especially for wintertime situations. Expanded representation of lower-voltage networks is highly desirable, particularly if the lower voltage networks include generation [3]. Sensitivity cases help determine appropriate modeling. Time domain (dynamic) simulation is required if the loads have fast dynamics. Dynamic simulation may be desired for confirmation of steady-state analysis when the load dynamics are slow.

5. Reactive power reserve should be monitored both in analyzing simulation results and in operating a real system. Generators, synchronous condensers, and static var compensators should not be in a control limit state following relatively high probability contingencies.

Example deterministic criteria. The load area is tested during heavy load conditions. A major reactive power source (generator, SVC, or shunt capacitor bank) is out of service prior to a large disturbance.

For a single outage (first contingency), the system must serve the predisturbance load with no fast-acting reactive power sources at control limits; a specified small amount of fast-acting reactive power reserve must remain for control of normal load fluctuations. Automatic local or centralized shunt reactive power compensation switching is allowed.

For credible multiple outages (e.g., outages of double circuit lines or lines with common terminations), undervoltage load shedding is allowed. Other emergency controls such as blocking of tap changers or direct load shedding are allowed. Consumer voltages must be above 90%. No additional power plants or transmission lines shall be in danger of tripping.

Performance is confirmed by the most realistic simulation methods available.

9.2 Solutions: Generation System

Improving voltage stability at the generation system involves planning, control and protection, and operation and maintenance.

Planning. The reliability benefits of generation in load areas should be a factor in generation siting decisions. This requires education of policy makers; remote generation alternatives mean either additional transmission or decreased power system reliability.

Gas turbines in load areas should be designed for fast start-up. Gas turbines and other expensive-to-operate generators could be designed to be de-clutched for operation as synchronous condensers [6,7].

For voltage stability improvement, several utilities are overhauling rather than retiring older synchronous condensers.

As discussed in Chapter 5 (Example 5-1), specifying lower power factor generation will increase fast-acting reactive reserves. If generators normally operate near unity power factor, the reduced generator losses will reduce the life-cycle cost increase of the larger, lower power factor machines.

New non-utility generation should be reviewed with regard to voltage stability performance [8].

Load tap changing on generator step-up (GSU) transformers and auxiliary system transformers have advantages for voltage stability. LTC transformers allow the transmission side voltage to be maintained at the highest possible value without regard to terminal voltage. For fixed tap transformers, voltage stability should be a factor in determining tap settings.

Excitation system control and protection. For transient voltage stability, modern high initial response and high ceiling excitation systems help induction motors reaccelerate following faults.

Excitation controls are discussed in detail in Chapter 5, Section 2. Voltage stability can be improved by better regulation of the generator high side voltage. This can be by line drop compensation or by an outer control loop. Keeping transmission voltage high minimizes the increase in reactive losses following a disturbance. Examples in Chapters 6 and 7 show the benefits. *This is one of the most cost-effective methods of improving voltage stability.*

Many older generators do not have overexcitation limiters, or do not have adequate overexcitation limiters. Overexcitation limiters that trip the excitation system to the predisturbance excitation level (typically near unity power factor) can cause disaster. If overexcitation limiters are not used, protection often trips the generator if the field current remains above rated current; this can also cause disaster. The excitation system should be modernized to incorporate continuously regulating field current limiters.

Active power controls. Fast generation changes can improve voltage stability. In attempting fast generation changes, we are in a race with load restoration by tap changing and generator current limiting. Figure 5-10 shows the overall concepts; Chapter 5, Section 4 describes details.

Protection. Undesirable operation of "transmission system backup protection" has been a factor in several voltage collapses. This stator-connected protection should be carefully reviewed—see Chapter 5, Section 2.

Maintenance. Generation systems are very complicated. Proper maintenance of both main equipment and control and protection is obviously important.

Verification and maximization of reactive power capability is especially important [9]. This includes control settings such as overexcitation limiters, alarm settings, protection, tap changer settings, and limitations imposed by auxiliaries. A testing program at Public Service Company of Colorado resulted in a 50% increase in reactive power capability from their generators. As described in Appendix F, the January 12, 1987 voltage collapse in Western France was greatly exasperated by the tripping of eight thermal plants due to field current protection defects.

Operation. To reduce power imports, gas turbines and other generators in load areas may have to be run out-of-merit (uneconomically) during high loading periods.

For loss of load-area generation contingencies, spinning reserve should be available in the load area. Generation control (AGC) should rapidly activate the spinning reserve.

In some cases, it may be advantageous to reduce real power loading of generators in load areas to allow higher reactive power loading. Power should be rescheduled over lightly loaded lines (Chapter 5, Section 4).

Reactive power output of generators should be closely monitored. The reactive power capability of generators should be known by control center operators. Shunt capacitor banks should be used to maintain fast-acting reactive power reserves at generators.

Power plant operators should be trained to allow high reactive power output of generators during infrequent emergency conditions. Automatic controls will reduce reactive power output as time-overload capability is used.

Modeling. For power flow programs, PV buses with simple Q_{max} and Q_{min} limits should be avoided. At a minimum, reactive power limits should be a function of active power loading as given by generator capability curves. Advanced programs directly calculate the field and armature currents to enforce limits [5]. Longer-term dynamics programs model excitation limiting control and protection. A challenge is to develop accurate data bases for the improved models—the effort should be made.

9.3 Solutions: Transmission System

Chapter 3 describes the transmission system including reactive power compensation.

Transmission design. For new construction, voltage stability is enhanced by high surge impedance loading values. This is accomplished by bundled conductors and compact line design [10]. The reduced reactance by increasing the number of subconductors per phase is equivalent to uniformly distributed series compensation (without subsynchronous resonance or other concerns).

New lines are difficult to build because of environmental, visual amenity, or financial reasons. If we have the opportunity to build a line, we should consider design for high capacity (high thermal capacity, low loss, and high surge impedance loading). Double circuit EHV lines should be considered.

An increase in the number of subconductors of a new high capacity line tends to optimize loadings of the new line and lower capacity parallel lines.

Example 9-2. Calculate the inductive reactance, charging power, and surge impedance loading for several 500-kV line designs. Similar to Example 3-1, the radius and *GMR* for all conductors are 2.5 and 2 cm, and the spacing between subconductors is 46 cm. Frequency is 60 Hz and operating voltage is 540 kV. Similar to tower types 5L8 and 5L2 from the EPRI Transmission Line Reference Book [11], the spacing between phases is 9.4 m. Calculate parameters for horizontal and delta phase configurations, and for 2, 3, and 4 subconductors per phase.

Solution: Equations 3.1 through 3.7 are used to develop the following table. Comparing the extreme cases (two subconductors/horizontal and four subconductors/delta), an increase in surge impedance of 29% percent is obtained. Using Equation 3.8, this is equivalent to 40% percent uniformly distributed series compensation.

Table for Example 9-2

	Horiz. 2 cond.	Horiz. 3 cond.	Horiz. 4 cond.	Delta 2 cond.	Delta 3 cond.	Delta 4 cond.
x_l,Ω/km	0.365	0.325	0.299	0.347	0.308	0.282
Q_{chg}, MVAr/km	1.29	1.44	1.56	1.36	1.52	1.66
P_0, MW	1017	1136	1234	1068	1201	1311

Additional subconductors may also be considered for reconductoring projects, perhaps for 230-kV lines as well as EHV lines. Utilities may, for example, have older 500-kV lines with conservative tower design and only two subconductors per phase. Additional subconductors could be added with only minor tower reinforcements.

Reactive power compensation. As described in Chapter 3, series capacitors have advantages over shunt capacitors and should be considered.

Long EHV transmission lines usually require shunt reactors. Switchable shunt reactors can be disconnected during voltage emergencies by operators, or by undervoltage relays. Tripping of 230-kV shunt reactors by undervoltage relays is used by Florida Power and Light Company—the controls were implemented after a series of voltage collapses in 1982 [12].

As discussed above, shunt capacitor banks can be used to allow generators to operate near unity power factor. This may require transmission system shunt capacitor banks rather than subtransmission and distribution shunt capacitors.

Mechanically switched shunt capacitors and static var compensators improve voltage stability. Application is discussed in Sections 3 and 4 of Chapter 3.

Bonneville Power Administration is developing a new low-cost method to improve the effectiveness of large 230-kV and 500-kV shunt capacitor banks [13]. During low voltage, series groups of wye-grounded shunt capacitors banks are temporary shorted to reduce capacitive reactance, and thereby increase reactive power output. Some of the temporary (10–30 minutes) overvoltage capability of capacitors is used. Voltage-controlled medium voltage switches are used to short the series groups. Figure 9-2 shows the concept for a 168 MVAr, 241.5-kV wye-grounded shunt capacitor bank. Figure 9-3 shows the modified capacitor bank characteristic. Shorting 3 of 14 series groups increases capacitor reactive power output by 14/11 or 127%.

Optimal power flow programs are useful for minimizing reactive power additions—subject to constraints and contingencies. Most programs, however, only satisfy voltage magnitude constraints and do not directly address voltage instability/collapse. Optimal power flow is an area of active development and improved software can be expected in the future. Replacement of PCB (polychlorinated biphenyl) capacitors should be optimized for the current system—with due regard for voltage stability.

Controls. Automatic on-load tap changing on large EHV/HV autotransformers can improve voltage stability. By regulating the voltage of the high voltage system, shunt capacitor output and high voltage system line charging is supported, and reactive losses are minimized. Tap changing at bulk power delivery substations and at distribution voltage regulators will not occur because of the faster regulation of the high voltage system. Voltage

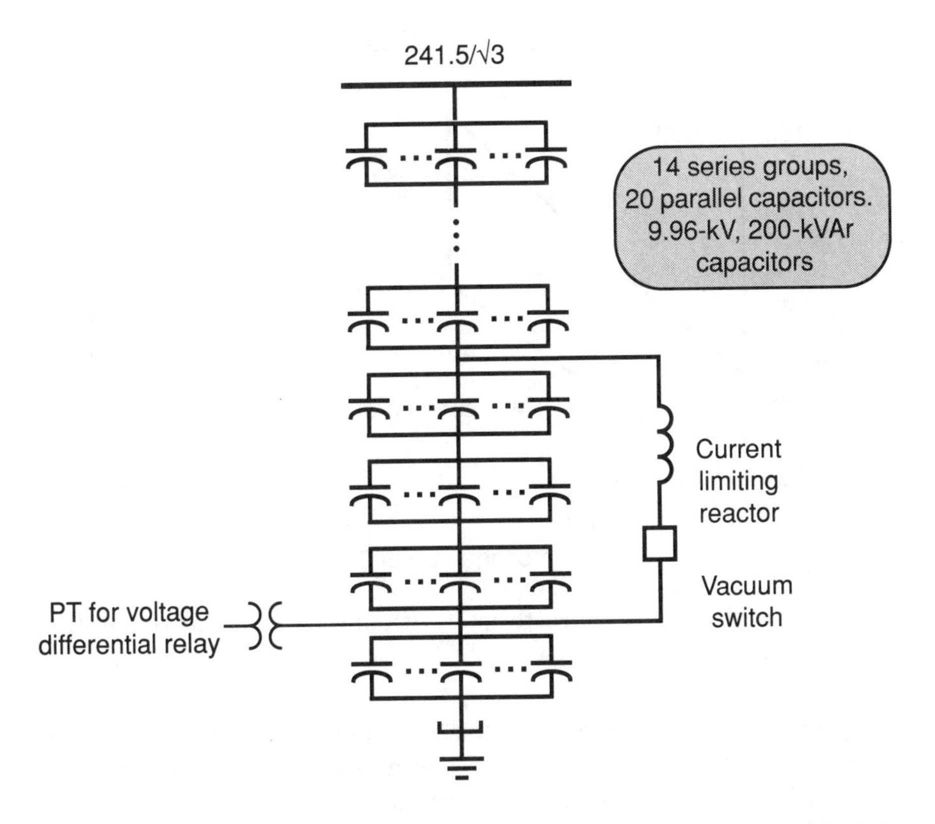

Fig. 9-2. Shorting of capacitor series groups for Bonneville Power Administration 168 MVAr shunt capacitor bank at Olympia Substation.

sensitive load will, however, be restored faster and undervoltage load shedding won't be as effective.

The tap changing will sag the EHV voltage. This can be compensated for by capacitor bank insertion, tripping of shunt reactors, and control of EHV-side voltage at generators. Tokyo Electric Power Company has developed a microprocessor-based controller for coordinated control of capacitor bank switching and network transformer tap changing [14].

As described in Chapter 7, Section 5, fast switching of shunt capacitor banks or shunt reactors following a disturbance can prevent the longer-term voltage instability mechanism from even starting. This may require use of remote signals.

Another control is automatic line reclosing following short circuit clearing. In contrast to high speed reclosing for transient stability, automatic line reclosing for the slower forms of voltage instability need not be as fast. For instance, ten seconds delay allows time for synchronizing oscillations

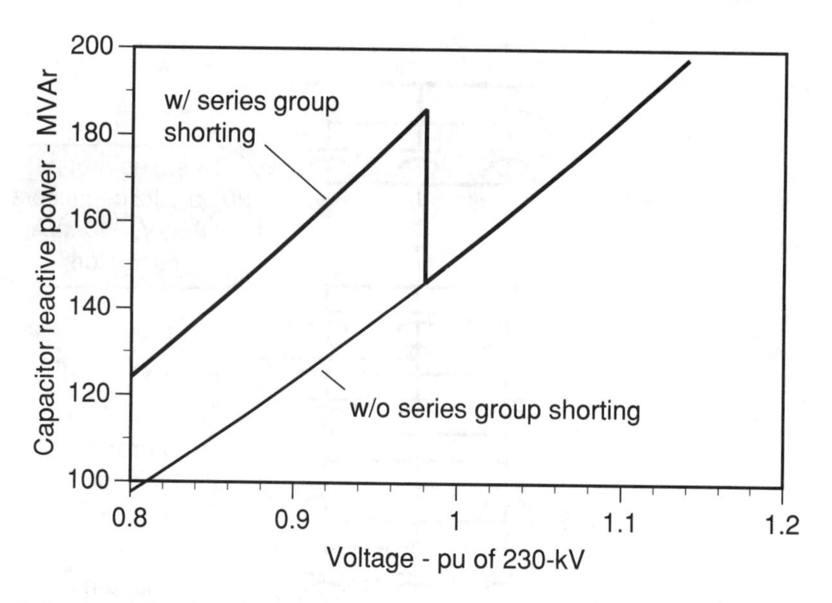

Fig. 9-3. Capacitor bank characteristic with and without shorting of three series groups at 0.98 pu voltage.

and generator torsional oscillations to damp out. The longer delay increases the chance for successful reclosing since more time is available for arc deionization. Automatic reclosing should be faster than capacitor switching, tap changing, or load shedding. Circuit breaker synchronizing check relays, if used, should not be set too restrictively.

Protective relaying. Many blackouts have been caused by protective relays operating on overload—their purpose is to operate for short circuits. On main grid lines, zone 3 impedance relays are usually the cause. With redundant relay sets, and with breaker failure relaying and bus protection (local backup), there is little need to apply zone 3 relays. Get rid of them. On subtransmission lines, overcurrent relays may be used in place of impedance relays. Make sure they won't operate on overload. Apparent impedance and line current should be monitored in simulation programs.

The December 19, 1978 French voltage collapse (Appendix F) was triggered by tripping of a critical 400-kV line by an overload relay. The line tripped with 20 minutes time delay after the overload relay operated and gave an alarm.

When an overloaded line is in danger of damage, real-time or dynamic line thermal rating equipment [15,16] can prevent unnecessary line tripping. The line capability is based on ambient conditions (temperature, wind, etc.). Due to the favorable ambient conditions, real-time line ratings are especially valuable for wintertime voltage stability situations.

Operation. During heavy load and emergency conditions, operators must use the means at their disposal (e.g., capacitor bank and reactor switching) to keep transmission voltages as high as allowed. During load buildup, capacitor banks must be applied early to "keep under the voltage."

HVDC transmission. HVDC power control should be used to improve voltage stability. In some cases, fast power increase is needed. For example, for the western North American interconnection shown on Figure 5-10, outage of the Pacific HVDC Intertie causes voltage stability problems in Northern California [17]. Fast power increase on the Intermountain Power Project would relieve overload on the Pacific AC Intertie, and improve voltage stability.

As described in Chapter 8, fast reduction of DC power is often required to release reactive power into the power system. Controls to reduce direct current and power for low ac voltage can improve voltage stability, as can inverter controls to regulate ac voltage.

9.4 Solutions: Distribution and Load Systems

Voltage stability is essentially load stability where the "load" is that seen from the bulk power system. This "load" includes the subtransmission and distribution systems. Effective solutions to voltage stability problems can be found at the problem source.

Planning, simulation studies. Upgrading subtransmission and distribution circuits, perhaps for energy conservation, will help voltage stability by reducing feeder impedances. Use of higher voltage distribution circuits will help.

Detailed representation of subtransmission systems is desirable for voltage stability studies. Representation may include equivalents for feeder impedances, representation of non-utility generation, and dynamic motor equivalents.

Capacitor banks. Shunt capacitor banks should usually be located on the regulated side of LTC voltage regulators. The shunt capacitor banks are then constant reactive power sources.

Control of voltage by switched shunt capacitors or series capacitors, rather than LTC transformers and distribution voltage regulators, will improve voltage stability.

Capacitor banks controlled by current, reactive power, time, or temperature should have backup, voltage-based control. Capacitor banks will then be inserted for bulk power system problems.

Tap changing. A particularly simple and often effective method to improve voltage stability is to block LTC transformer tap changing for low unregulated side (transmission side) voltage. This is most effective at substations serving high power factor loads; for highly shunt compensated loads, benefits may be small or negative. If the load is some distance from the LTC transformer, tap changer blocking may also be counterproductive.

Ontario Hydro has implemented tap changer blocking on fourteen transformer stations in the Ottawa area [4, page 178]. Tap changers are blocked when the high side voltage (230-kV or 115-kV) drops below a set value for a specified time. Tap changing is unblocked when voltage has recovered to stable values for a specified time. The tap changer blocking controls are coordinated with automatic reclosing, capacitor switching, and load shedding.

Blocking tap changers to allow distribution voltage to fall will result in loss of load diversity as thermostats regulate constant energy loads. This could be significant in wintertime voltage stability situations. Blocking tap changers would, however, allow time for other actions.

Blocking tap changers may be less cost-effective if numerous distribution voltage regulators are used rather than large LTC bulk power delivery transformers. An alternative is to allow tap changing only one or two boost steps above the value normally reached during heavy load conditions. Another alternative is to use long intentional time delays between individual tap steps as discussed in Chapter 4, Section 4. This allows more time for corrective action. Wider tap changer bandwidth settings could also be considered.

If two or more tap changer transformers or regulators are in series, operation should be coordinated so that the tap changing closer to the load has a much longer time delay.

Voltage reduction. As described in Chapter 4, voltage reduction during critical conditions is widely used to obtain load relief. For voltage stability, the reduction in reactive power load may be particularly significant. Because of distribution transformers operating in saturation, a 1% reduction in voltage may cause a sustained reduction in reactive power load by 4–7%.

Undervoltage load shedding. Undervoltage load shedding [18–19] is a cost-effective, decentralized voltage stability solution for infrequent disturbances. It's a valuable backup for primary solutions to voltage stability problems. Undervoltage load shedding removes a burden from system operators who otherwise might be required to rapidly shed load manually during emergencies. It allows operators to take more risk in achieving economic system operation. The time delay must be short (1–1.5 seconds) to prevent stalling of induction motors during the final phase of voltage instability [18]. If most of the load is static and highly voltage sensitive (heating and lighting), the time delays may be longer. The undervoltage relays should respond to balanced, positive sequence voltage drops, or be blocked from operating for unbalanced conditions.

Several utilities have implemented undervoltage load shedding programs [20]. Reference 19 describes a system installed in 1981 by the Tennessee Valley Authority to protect against transient voltage instability in an area with high air conditioning load. The relays, installed at nine 161-kV substations, have time delays between 60 and 105 cycles. During the summer of 1987, the system prevented voltage collapse on three occasions.

Chapter 7, Section 4 describes undervoltage load shedding installed in the Puget Sound area of the Pacific Northwest to prevent longer-term voltage collapse. The loads are highly voltage sensitive and the time delays are not critical. Similar undervoltage load shedding is planned for the Portland load area.

Direct load tripping. Direct detection of major contingencies can initiate tripping of load. In the Puget Sound area, for example, outage of 500-kV cross-Cascade Mountain 500-kV lines during heavy load conditions initiates tripping of aluminum reduction plant load. This system, however, will be phased out with the installation of undervoltage load shedding and other system additions.

Florida Power and Light Company had developed the Fast Acting Load Shedding (FALS) program which runs at their system control center [21]. The system differentiates between generation loss below and above approximately 1200 MW, and initiates 800 MW of load shedding for generation loss above 1200 MW. The load shedding occurs about twenty seconds after the outage. The load is shed before load restoration by tap changing and before field current limiting at generators. The time delay allows automatic switching of transmission shunt capacitor banks; this may eliminate the need for load tripping. There is a similar system for transmission corridor outages called Corridor Fast Acting Load Shed (CFALS).

Direct load control and distribution automation. We would like to shed load less disruptively than by direct load tripping or undervoltage load shedding, [18,22]. Rapidly turning off air conditioners, water heaters, electric heating, or other load for five to twenty minutes during an emergency is attractive. For longer-term voltage stability, extremely fast action is not required—tens of seconds or minutes are available. The controls provide load relief, and the time needed to start gas turbines or reschedule generation. One concept is for the utility to communicate emergency conditions to consumer microprocessors by a large increase in the current cost of electricity [23]. Thermostat settings or load demand would then be changed to reduce consumption.

Many utilities have implemented or are considering direct load control which is a type of demand side management. The requirements for emergency load shedding are sensitive methods to detect impeding voltage collapse, and fast communications and actuators. The technology for fast load relief is available today [24], but widespread use of fiber optics and comprehensive communication systems such as ISDN (Integrated Services Digital Networks) [25] will improve feasibility in the near future—at least for industrial and commercial loads.

Direct load control can be initiated based on activation of reactive power reserve at generators and static var compensators. As previously discussed, reactive power reserve activation is a sensitive indicator of impeding voltage instability.

Direct load control can also improve voltage stability through switched capacitor bank control, tap changer control, and voltage reduction. Future coordinated distribution network and transmission network control can aid bulk power system problems.

Direct load control can be used for inexpensive testing of load characteristics. Capacitor banks can be switched off by SCADA to lower voltage, and the resulting active and reactive power response of the loads can be monitored and analyzed at control centers. Voltage reduction controls can be used for the same purpose.

9.5 Power System Operation

After system planners, system engineers, design engineers, and control engineers have developed a power system—probably with a plethora of emergency controls—the burden falls on system operators to balance system reliability against economics. Voltage security problems greatly adds to the burden. Automation and computers, while relieving much of the burden, increases the complexity of power system operation.

Referring to our classification of voltage stability shown on Figure 2-1, it's clear that operators cannot act fast enough for transient voltage stability. For longer-term voltage stability, it's questionable whether operators should be asked to perform corrective control in the first several minutes following a major disturbance. The operator's role is rather to ensure that the power system predisturbance state is secure for the most probable disturbances. For longer-term voltage stability involving load buildup, however, operators have an important role.

In keeping a secure predisturbance state, operators may have to reschedule generation, switch capacitor banks, or order voltage reductions. During an abnormally large load buildup, or during insecure operation due to equipment outages, operator may need to shed load.

For the generation system and for the transmission system, we have already discussed some operational aspects.

Energy Management Systems. In order to make critical decisions, operators need the best possible information. Energy Management Systems (EMS) provide a variety of measured and computed data. State estimation methods are commonly used to filter measured data to provide the network static "state" consisting of voltage magnitudes and angles. State estimation is valuable in that power flow model inaccuracies, particularly involving reactive power flow, are resolved. In comparing off-line power flows with real system power flow, difficulty in matching reactive power flows are often encountered, implying model errors for voltage stability studies.

State estimation provides the system model for security assessment software [26]. Voltage security software is being developed and will be an important part of energy management systems.

For a given operating condition, possibly with unusual combinations of outages and power flows, fast contingency screening and ranking is often desired. Reference 27 describes some methods for voltage security. Contingency screening and ranking is followed by more detailed analysis of the most critical contingencies.

Various computer methods have been developed to assist the operator in "reactive power management" and voltage control. Most methods involve optimization of economics (transmission system active power loss minimization) subject to voltage magnitude constraints [28]. "Security-constrained optimization" ensures voltage magnitude and other constraints are met for first contingencies. At present, their value for voltage stability/security monitoring is limited since, for example, they usually don't include voltage stability constraints such as requirements for reactive power reserve. Development of new software, however, is underway.

Artificial intelligence is another approach to centralized reactive power and voltage control. For example, we have stressed the need to apply capacitor banks so that generators operate near unity power factor; an expert system could assist operators. For voltage stability, the University of Liège and Electricité de France are investigating a decision tree method [29]. Many expert system methods are being developed [30].

Specific to voltage stability, especially longer-term voltage stability, several utilities have EMS functions to guide operator actions. Because of time constraints, static (power flow) analysis methods and artificial intelligence methods will likely predominate for at least the near future.

To date, computed *P–V* curves are the most widely used method of estimating voltage security, providing megawatt margin type indices.

Reference 31 describes the voltage security monitoring and voltage security assessment system implemented in February 1990 by Tokyo Electric Power Company. On-line voltage security monitoring runs at intervals of one minute and displays *P–V* curves for conditions one minute before, for the present time, and for ten minutes in the future. Reactive power reserves are displayed for the same times, along with guidance for control. The voltage security assessment runs contingencies cases in the morning for the daily peak load conditions.

References 32 and 33 describe voltage security monitoring systems developed, respectively, by the National Grid Company (U. K.) and Electricité de France. Both systems determine the distance to the maximum loadability point (*P–V* curve critical point) by the extended sensitivity computation approach developed by Flatabø [34]. At least for the EdF system, emphasis is on longer-term voltage instability due to load buildup.

A different approach is used in the Belgium national control center. A fast optimization method is used to determine the maximum reactive power load of an area (nose of *Q–V* curve for a load area) [35]. The reactive power margin between the operating point and the maximum reactive power is an indication of voltage security or robustness for longer-term voltage stability involving large disturbances.

Post-disturbance MW or MVAr margins should be translated to predisturbance operating limits that operators can monitor.

Voltage stability indices are categorized and further described in Appendix B.

Besides software application programs, direct monitoring of key voltage security indicators may be very effective. The EMS could provide operator displays of generator capability curves, with indication of current operating point and the target operating area. As preventive control, operators would switch reactive power compensation to maintain fast-acting reactive power reserves at generators, synchronous condensers, and SVCs. Real-time updating of generator capability curves may be desirable in some cases.

Centralized emergency controls. For longer-term voltage stability, control actions sometimes must be taken within a minute following the disturbance. A strong argument can be made for decentralized local controls such as generator regulation of high-side voltage, voltage switched capacitor banks and shunt reactors, local voltage (unregulated-side) based tap changer blocking, and undervoltage load shedding.

Nevertheless, there are reasons for dispatch center based centralized controls. Since there is insufficient time for algorithmic-based computations, controls must be pre-programmed based on off-line studies. Several artificial intelligence approaches have been proposed. Lachs, et al. [36] propose a relatively simple expert system approach. Van Cutsem, et al. [37] propose a decision tree approach. In both methods, emergency actions are taken based mainly on measured change of generator and SVC reactive power, and change of voltage magnitudes. Among other actions, artificial intelligence could be used to trigger distribution automation actions.

As shown in Chapter 7, Section 5, centralized automatic voltage control can be more sensitive to disturbances than purely local voltage magnitude based control.

In addition to reactive power and voltage measurements, control can be based on direct detection of major outages. The Florida Power and Light FALS system described above is one example.

EMS software for voltage stability and centralized emergency controls for voltage stability are areas of high current interest, with considerable research and development in progress.

Training. Both control center and power plant operators should be trained in the basics of voltage stability. They should know how to recognize voltage emergencies, and know what emergency actions may have to be taken. Operator actions for voltage stability are an extension of voltage and reactive power control actions for normal conditions.

Training ideally should include operator training simulator sessions. References 38–40 describe the recently developed EPRI training simulator. This is a real-time dynamic model of the power system that interfaces with utility energy management system controls such as automatic generation controls. Only the slower uniform system frequency dynamics are represented. Reference 41 describes use of the simulator at Philadelphia Electric Company for restoration following a voltage collapse.

9.6 Summary: the Voltage Stability Challenge

Voltage stability is likely to challenge utility planners and operators for the foreseeable future. As load grows, and as new transmission and load-area generation becomes increasingly difficult to build, more and more utilities will face the voltage stability challenge. The relatively recent problem of solar magnetic disturbances adds another dimension that is important for many utilities.*

Fortunately, many creative people are working on new analysis methods and on innovative solutions to the voltage stability challenge. The subject is being approached from many viewpoints. There is need, however, to involve others, such as distribution automation engineers. Voltage stability is essentially load stability, and many of the cost-effective solutions involve load control.

References

1. M. V. F. Pereira and N. J. Balu, "Composite Generation/Transmission Reliability Evaluation," *Proceedings of the IEEE*, Vol. 80, No. 4, pp. 470–491, April 1992.
2. CIGRÉ Working Group 38.01, "Planning Against Voltage Collapse," *Electra*, pp. 55–75, March 1987.
3. J. D. McCalley, J. F. Dorsey, Z. Qu, J. F. Luini, and J. L. Filippi, "A New Methodology for Determining Transmission Capacity Margin in Electric Power Systems," *IEEE Transactions on Power Systems*, Vol. 6, No. 3, pp. 944–951, August 1991.
4. IEEE Committee Report, *Voltage Stability of Power Systems: Concepts, Analytical Tools, and Industry Experience*, IEEE publication 90TH0358-2-PWR, 1990.
5. B. Gao, G. K. Morison, and P. Kundur, "Voltage Stability Evaluation Using Modal Analysis," *IEEE Transactions on Power Systems*, Vol. 7, No. 4, pp. 1529–1542, November 1992.
6. W. B. Jervis, J. G. P. Scott, and H. Griffiths, "Future Application of Reactive Compensation Plant on the CEGB System to Improve Transmission Network Capability," CIGRÉ, *Proceedings of 33rd Session*, Vol. II, paper 38-06, 1988.
7. F. Iliceto, E. Cinieri, F. Gatta, and A. Erkan, "Optimal Use of Reactive Power Resources for Voltage Control in Long Distance EHV Transmission: Applications to the Turkish 420-kV System," CIGRÉ, *Proceedings of 33rd Session*, Vol. II, paper 38-03, 1988.
8. H. Kirkham and R. Das, "Effects of Voltage Control in Utility Interactive Dispersed Storage and Generation Systems," *IEEE Transactions on Power Apparatus and Systems*, Vol. PAS-103, No. 8, pp. 2277–2282, August 1984.
9. P. B. Johnson, S. L. Ridenbaugh, R. D. Bednarz, and K. G. Henry, "Maximizing the Reactive Capability of AEP Generating Units," *Proceedings of American Power Conference*, pp. 373–377, April 1990.
10. *Compacting Overhead Transmission Lines*, Proceedings of CIGRÉ Symposium, Leningrad, 3–5 June 1991.
11. Electric Power Research Institute, *Transmission Line Reference Book, 345 kV and Above*, second edition, 1982 (prepared by General Electric Company).

---

*Geomagnetic storms cause quasi-dc currents to flow in a path which includes grounded transformers and transmission lines. Transformer saturation due to the quasi-dc current results in low voltages and harmonic generation. Shunt capacitors banks are harmonic current sinks, and may overload and trip.

12. IEEE Committee Report, "VAR Management—Problem Recognition and Control," *IEEE Transactions on Power Apparatus and Systems*, Vol. PAS-103, No. 8, pp. 2108–2116, August 1984.
13. C. W. Taylor, "A Novel Method to Improve Voltage Stability using Shunt Capacitors," *Proceedings: Bulk Power System Voltage Phenomena II—Voltage Stability and Security*, Deep Creek Lake, Maryland, pp. 395–401, 4–7 August 1991.
14. S. Koishikawa, S. Ohsaka, M. Suzuki, T. Michigami, and M. Akimoto, "Adaptive Control of Reactive Power Supply Enhancing Voltage Stability of a Bulk Power Transmission System and a New Scheme of Monitor on Voltage Security," *CIGRÉ*, paper 38/39-01, 1990.
15. R. A. Moraio, S. D. Foss, and H. Schraushuen, "Thermal Line Rating for Improved Transmission System Utilization," *Power Technology International 1991*, pp. 41–48.
16. L. Cibulka, W. J. Steeley, and A. K. Deb, "PG&E's ATLAS (Ambient Temperature Line Ampacity System) Transmission Line Dynamic Thermal Rating System," *CIGRÉ*, paper 22-102, 1992.
17. W. Mittelstadt, C. Taylor, M. Klinger, J. Luini, J. McCalley, and J. Mechenbier, "Voltage Instability Modeling and Solutions as Applied to the Pacific Intertie," *CIGRÉ*, paper 38-230, 1990.
18. C. W. Taylor, "Concepts of Undervoltage Load Shedding for Voltage Stability," *IEEE Transactions on Power Delivery*, Vol. 7, No. 2, pp. 480–488, April 1992.
19. H. M. Shuh and J. R. Cowan, "Undervoltage Load Shedding An Ultimate Application for Voltage Collapse," *Georgia Institute of Technology 46th Annual Protective Relaying Conference*, April 29–May 1, 1992.
20. IEEE Power System Relaying Committee Working Group K12, "System Protection and Voltage Stability," Draft 5, September 1992.
21. S. A. Nirenberg, D. A. McInnis, and K. D. Sparks, "Fast Acting Load Shedding," *IEEE Transactions on Power Systems*, Vol. 7, No. 2, pp. 873–877, May 1992.
22. C. W. Taylor, "Solving Bulk Transmission System Security Problems with Distribution Automation," *Proceedings, Third International Symposium on Distribution Automation and Demand Side Management*," January 11–13, 1993, Palm Springs, California.
23. M. G. Adamiak, D. C. Roberts, and S. D. Ketz, "A Microprocessor-Based System for the Implementation of Variable Spot Pricing of Electricity," *IEEE Computer Applications in Power*, pp. 43–48, October 1990.
24. H. E. Caldwell, Jr., "Load Management Reaches for the Stars," *Transmission and Distribution*, pp. 58–62, February 1991.
25. Special issue on Integrated Services Digital Networks, *Proceedings of the IEEE*, Vol. 79, No. 2, February 1991.
26. N. Balu, T. Bertram, A. Bose, V. Brandwajn, G. Cauley, D. Curtice, A. Fouad, L. Fink, M. G. Lauby, B. F. Wollenberg, and J. N. Wrubel, "On-Line Power System Security Analysis, *Proceedings of the IEEE*, Vol. 80, No. 2, pp. 262–280, February 1992.
27. G. C. Ejebe, H. P. Van Meeteren, and B. F. Wollenberg, "Fast Contingency Screening and Evaluation for Voltage Security Analysis," *IEEE Transactions on Power Systems*, Vol. 3, No. 4, pp. 1582–1590, November 1988.
28. O. Alsac, J. Bright, M. Prais, and B. Stott, "Further Developments in LP-Based Optimal Power Flow," *IEEE Transactions on Power Systems*, Vol. 5, No. 3, pp. 697–711, August 1990.

29. L. Wehenkel, T. Van Cutsem, B. Heilbronn, and M. Goubin, "Decision Trees for Preventive Voltage Stability Assessment," *Proceedings: Bulk Power System Voltage Phenomena II—Voltage Stability and Security*, Deep Creek Lake, Maryland, pp. 229–240, 4–7 August 1991.
30. CIGRÉ Task Force 38-06-01, *Expert Systems Applied to Voltage & VAr Control*, 1991.
31. M. Suzuki, S. Wada, M. Sato, T. Asano, and Y. Kudo, "Newly Developed Voltage Security Monitoring System," *IEEE Transactions on Power Systems*, Vol. 7, No. 3, pp. 965–973, August 1992.
32. A. O. Ekwue, R. M. Dunnett, N. T. Hawkins, and W. D. Laing, "On-Line Voltage Collapse Monitor," *Proceedings: Bulk Power System Voltage Phenomena II—Voltage Stability and Security*, Deep Creek Lake, Maryland, pp. 253–255, 4–7 August 1991.
33. C. Lemaître, J. P. Paul, J. M. Tesseron, Y. Harmand, and Y. S. Zhou, "An Indicator of the Risk of Voltage Profile Instability for Real-Time Control Applications," *IEEE Transactions on Power Systems*, Vol. 5, No. 1, pp. 154–161, February 1990.
34. N. Flatabø, R. Ognedal, and T. Carlsen, "Voltage Stability Condition in a Power Transmission System Calculated by Sensitivity Methods," *IEEE Transactions on Power Systems*, Vol. 5, No. 4, pp. 1286–1293, November 1990.
35. T. Van Cutsem, "A Method to Compute Reactive Power Margins with Respect to Voltage Collapse," *IEEE Transactions on Power Systems*, Vol. PWRS-6, No. 2, pp. 145–156, February 1991.
36. W. Lachs, Y. L. Zhou, D. Stannite, and I. F. Morrison, "Automatic Control of System Voltage Stability by an Expert System," *Proceedings of Tenth Power System Computing Conference*, Graz, Austria, pp. 1057–1064, Butterworths, London, 1990.
37. T. Van Cutsem, L. Wehenkel, M. Pluvial, B. Heilbronn, and M. Goubin, "Decision Trees for Detecting Emergency Voltage Conditions," *Proceedings: Bulk Power System Voltage Phenomena II—Voltage Stability and Security*, Deep Creek Lake, Maryland, pp. 217–228, 4–7 August 1991.
38. *Operator Training Simulator*, EPRI Final Report EL-7244, May 1991, prepared by EMPROS Systems International.
39. M. Prais, G. Zhang, A. Bose, and D. Curtice, "Operator Training Simulator: Algorithms and Test Results," *IEEE Transactions on Power Systems*, Vol. 4, No. 3, pp. 1154–1159, August 1989.
40. M. Prais, C. Johnson, A. Bose, and D. Curtice, "Operator Training Simulator: Component Models," *IEEE Transactions on Power Systems*, Vol. 4, No. 3, pp. 1160–1166, August 1989.
41. R. F. Chu, E. J. Dobrowolski, E. J. Barr, J. McGeehan, D. Scheurer, and K. Nodehi, "Restoration Simulator Prepares Operators for Major Blackouts," *IEEE Computer Applications in Power*, Vol. 4, No. 4, pp.46–51, October 1991.

Appendix A

Notes on the Per Unit System

In everyday life, *per cent* is widely used to increase understanding. For example, on August 19, 1991 following the short-lived coup d'état in the Soviet Union, the Tokyo (Nikkei) stock market index dropped 5.95%. The 5.95% drop corresponded to 1357.61 points—a number which is meaningless to most of us.

Per unit is a percentage value divided by 100.

In engineering, scaling or normalizing of physical values is often useful. In power system analysis, a per unit (pu) system is used to express a physical variable as a fraction of a base or reference value. The base value is usually a rated or full-load value. We state:

$$\text{pu value} = \frac{\text{value in physical units}}{\text{base value in same physical units}}$$

Per unit is commonly used for voltage, current, impedance, and power. In power flow and related calculations or computation, base apparent power is invariant. Base voltage varies because of transformation between several voltage levels. Base current and impedance at a point in the network are calculated once base power and voltage are specified.

For three-phase systems, the methods and formulas are provided in several books on power system analysis. For reference, we provide the following formulas. The normal conventions of three-phase power and line-to-line voltage are used.

$$\text{Base current from } S_{\text{base}} = \sqrt{3} V_{\text{base}} I_{\text{base}}$$

$$Z_{\text{base}} = \frac{V_{\text{base}}^2}{S_{\text{base}}}$$

To convert from one impedance base to another, the formula is:

$$Z_{\text{base new}} = Z_{\text{base old}} \left[\frac{V_{\text{base old}}}{V_{\text{base new}}} \right]^2 \left[\frac{S_{\text{base new}}}{S_{\text{base old}}} \right]$$

The base apparent power depends on the problem at hand. Examples are:

- When dealing with a single piece of equipment such as a generator, motor, transformer, or SVC, the equipment MVA (or MVAr) rating is used. The same base is used when the equipment is connected to a large power system, e.g., the one machine to infinite bus problem.
- The system short circuit capacity may be the base. Dividing the short circuit capacity by the equipment rating gives the short circuit ratio. The inverse of the short circuit ratio is the thévenin reactance in per unit of the equipment rating.
- For calculations involving a transmission line, the surge impedance loading is most appropriate. The base impedance is the surge impedance.
- For calculations or computer programs involving a power system with several or many components, the base MVA must, unfortunately, be chosen arbitrarily. The commonly used 100 MVA base is appropriate for lower voltage parts of a system, but inappropriate for EHV parts of the system (the surge impedance loading of 500-kV lines is around 1000 MW.)

Advantages of the per unit system are:

- Voltages have the same range in per unit (i.e., 1 ± 0.05 pu) in all parts of a system from EHV to distribution and utilization.
- When expressed in per unit, apparatus parameters usually fall in fairly narrow range regardless of apparatus size. For example, generator reactances in per unit are similar for both 100 MVA machines and 1000 MVA machines. This facilitates data checking and hand calculations.
- Analysis and computation for synchronous machines is greatly facilitated. Selection of base rotor quantities is quite involved. The companion book *Power System Stability and Control* describes per unit representation of synchronous machine.
- Ideal transformers with nominal turns ratios are eliminated by the per unit system.
- The $\sqrt{3}$ factor in three-phase circuit calculations is eliminated.

There are some disadvantages of the per unit system and some circumstances where the system is not needed.

- For transmission lines, it's the values of impedances and admittances in physical units (e.g., ohms/km) that are of the same magnitude regardless of voltage level or MVA rating.

- For dc transmission, there is no need for converting to per unit.
- The equation for three-phase power is not the same in per unit as in physical units. The engineer tends to forget about $\sqrt{3}$ factors when using physical units.
- Other normalization is sometimes better. For example, the inertia of rotating machines is normalized by dividing by the machine MVA rating. The units of the resulting inertia constant, H, are MW-seconds/MVA or seconds.

Appendix B

Voltage Stability and the Power Flow Problem

The power flow (load flow) problem is at the heart of power system analysis. Similar network solution techniques are used for steady state programs (power flow program, short circuit program) and dynamic programs. The power flow problem is very closely associated with voltage stability analysis.

Much of the literature and research on voltage stability deals with power flow computation methods. This appendix summarizes, for non-programmers, basics of this specialized subject. Books on power system analysis introduce the subject; Debs [1], Arrillaga and Arnold [2], and Dommel [3] provide detailed descriptions. We provide additional references.

The power flow problem solves the complex matrix equation

$$\mathbf{YV} = \mathbf{I} = \frac{\mathbf{S}^*}{\mathbf{V}^*} \tag{B.1}$$

$\mathbf{Y}$ is the network nodal admittance matrix, $\mathbf{V}$ is the unknown complex node voltage vector, $\mathbf{I}$ is the nodal current injection vector, and $\mathbf{S} = \mathbf{P} + \mathrm{j}\mathbf{Q}$ is the apparent power nodal injection vector representing specified load and generation at nodes. Equation B.1 is supplemented by equations for area interchange control or AGC, generator reactive power limits, bus voltage control, tap changer control, HVDC links, static var compensators, etc. These equations are sometimes called the control equations.

The Newton-Raphson method and fast decoupled methods are the two main ways of solving the power flow problem.

B.1 The Nodal Admittance Matrix

We assume balanced three-phase operation, allowing "per phase" or positive sequence representation. The nodal admittance matrix models the network. The diagonal terms, Y_{kk}, are the sum of all admittances connected to node k, including admittances to ground. Impedance loads are included in Y_{kk}. The off-diagonal terms, Y_{km}, are the negated sum of admittances between nodes k and m.

Note that off-diagonal terms, Y_{km}, are non-zero only if there is a direct connection between nodes k and m. In large power systems, substation busses (nodes) are directly connected by transmission lines or transformers to only a few other busses. Therefore, the nodal admittance matrix is extremely sparse—most of the off-diagonal terms are zeros. In large-scale programs, coding to exploit sparsity is vital. The zero terms are not stored, nor used in computations.

Example B-1. Calculate the nodal admittance matrix elements for the three bus system of Figure 6-1 (see Figure B-1). The load has a resistive component. Use a 100 MVA base and assume a 238.7/25-kV turns ratio for the transformer between nodes 2 and 3.

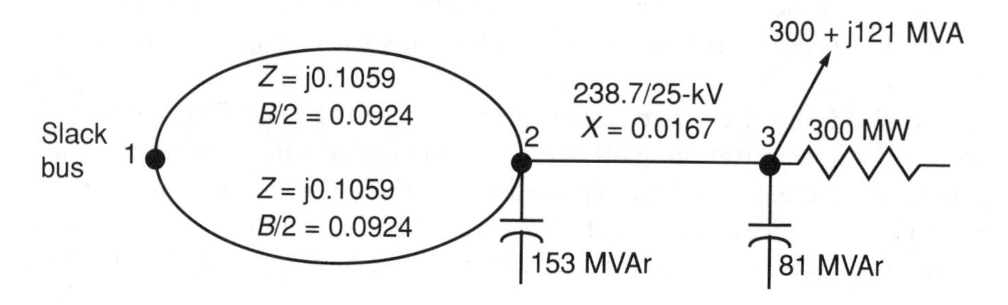

Fig. B-1. Three bus network. See Figure 6-1.

Solution: The transformer can be represented as a pi equivalent using the model given in Chapter 4. The off-nominal turns ratio is n = 230/238.7 = 0.9636. The matrix elements are:

$$Y_{11} = j2 \times (0.0924 - 1/0.1059) = \text{-j}18.70094$$

$$Y_{12} = Y_{21} = -j2 \times (-1/0.1059) = j\,18.88574$$

$$Y_{13} = Y_{31} = 0$$

$$Y_{22} = Y_{11} + j\,[(n\,(n-1)/0.0167\,) - n/0.0167 + 1.53] = -j72.76927$$

$$Y_{23} = Y_{32} = j\,[\,(n/0.0167)\,] = j57.69776$$

$$Y_{33} = 3.0 + j\,[0.81 - (n/0.0167\,) - (1-n)/0.0167]$$
$$= 3.0 - j59.07024$$

Bus types. Three basic bus or node types are used:

- *PQ bus*. A PQ bus is a load bus where the complex (active and reactive) load is specified. The specified load is usually constant, reflecting regulation of the load or load voltage. The complex node voltage is unknown.
- *PV bus*. A PV bus is a generator bus where active power and voltage magnitude is specified. In the real world, active power and voltage magnitude are kept constant by generator controls. The voltage angle at the generator terminals is the unknown. In production-grade programs, the voltage of a remote (e.g. high-side) bus can be constant rather than terminal voltage. The bus is subject to generator reactive power limits, and is traditionally converted to a PQ bus when limits are reached.
- *Slack bus*. One bus must be selected as a reference (voltage angle of zero). Also the difference between estimated and computed losses must be assigned to one or more busses. These two functions are usually assigned to a single large generator bus known as the slack bus. A slack bus is an infinite bus with constant voltage and unlimited real and reactive power capability. Both the voltage magnitude and angle are thus known at this bus.

Two equations are required for a PQ bus with voltage magnitude and angle as unknowns. One equation is required for a PV bus with voltage angle as unknown. No equation is needed for the slack bus.

B.2 The Newton-Raphson method

The most general and reliable algorithm to solve the power flow program is the Newton-Raphson method. It's a multi-variable formulation of Newton's method studied in calculus courses. The method involves iteration based on successive linearization using the first term of a Taylor expansion of the equations to be solved. From Equation B.1, we can write the equation for node k (row k) as:

$$\overline{I}_k = \sum_{m=1}^{n} \overline{Y}_{km}\overline{V}_m \tag{B.2}$$

$$P_k - jQ_k = \overline{V}_k^{*}\overline{I}_k = \overline{V}_k^{*}\sum_{m=1}^{n} \overline{Y}_{km}\overline{V}_m \tag{B.3}$$

or

$$P_k + jQ_k =$$

$$\sum_{m=1}^{n} Y_{km}V_kV_m[\cos(\theta_{km} - \gamma_{km}) + j\sin(\theta_{km} - \gamma_{km})] \tag{B.4}$$

Where $\theta_{km} = \theta_k - \theta_m$ and $\gamma_{km} = \arctan(B_{km}/G_{km})$.

We must now distinguish between specified or scheduled powers and powers calculated using Equation B.3 or B.4. The difference is the mismatch which becomes small as convergence of the iterative process is reached. The equations are:

$$\begin{aligned} \Delta P_k &= P_k^{Specified} - P_k \\ \Delta Q_k &= Q_k^{Specified} - Q_k \end{aligned} \tag{B.5}$$

The Newton-Raphson method solves the partitioned matrix equation:

$$\mathbf{J}\begin{bmatrix} \Delta\theta \\ \Delta\mathbf{V} \end{bmatrix} = \begin{bmatrix} \Delta\mathbf{P} \\ \Delta\mathbf{Q} \end{bmatrix} \tag{B.6}$$

where $\Delta\mathbf{P}$ and $\Delta\mathbf{Q}$ are mismatch vectors, $\Delta\mathbf{V}$ is the unknown voltage magnitude correction vector, and $\Delta\theta$ is the unknown voltage angle correction vector. $\mathbf{J}$ is the Jacobian matrix of partial derivative terms calculated analytically from Equation B.3. In Equation B.4, the voltages are in polar form, which is most common for power flow.

Jacobian terms. To avoid computation with trigonometric terms, rectangular coordinates are used even though the variables are in polar coordinates. The following definitions are used:

$$\overline{Y}_{km} = G_{km} + jB_{km}$$

$$\bar{V}_k = e_k + \mathrm{j}f_k$$

$$a_{km} + \mathrm{j}\,b_{km} = (e_m + \mathrm{j}f_m)\,(G_{km} + \mathrm{j}B_{km})$$

$$= (G_{km}e_m - B_{km}f_m) + \mathrm{j}\,(B_{km}e_m + G_{km}f_m) \tag{B.7}$$

Eight partial derivative terms are needed—four off-diagonal terms and four diagonal terms. The voltage magnitude correction terms are slightly modified to make the equations simpler; compensating terms are in the voltage magnitude correction vector. The partial derivative terms are [2,3,4]:

$$\frac{\partial P_k}{\partial \theta_m} = V_k V_m\,(G_{km}\sin\theta_{km} - B_{km}\cos\theta_{km}) = a_{km}f_k - b_{km}e_k \tag{B.8}$$

$$V_m\frac{\partial P_k}{\partial V_m} = V_k V_m\,(G_{km}\cos\theta_{km} + B_{km}\sin\theta_{km}) = a_{km}e_k + b_{km}f_k \tag{B.9}$$

$$\frac{\partial Q_k}{\partial \theta_m} = -V_m\frac{\partial P_k}{\partial V_m} \tag{B.10}$$

$$V_m\frac{\partial Q_k}{\partial V_m} = \frac{\partial P_k}{\partial \theta_m} \tag{B.11}$$

$$\frac{\partial P_k}{\partial \theta_k} = -Q_k - B_{kk}V_k^2 \tag{B.12}$$

$$V_k\frac{\partial P_k}{\partial V_k} = P_k + G_{kk}V_k^2 \tag{B.13}$$

$$\frac{\partial Q_k}{\partial \theta_k} = P_k - G_{kk}V_k^2 \tag{B.14}$$

$$V_k\frac{\partial Q_k}{\partial V_k} = Q_k - B_{kk}V_k^2 \tag{B.15}$$

Example B-2. Corresponding to Equation B.6, write the matrix equation for the three bus system shown of Example B-1. Assume node 2 is regulated by an SVC modeled as a PV bus.

Solution: We have one PV bus and one PQ bus resulting in three unknowns. There is no equation for the slack bus; the slack bus complex voltage is used in computing the mismatch powers. The equation is given below.

$$\begin{bmatrix} \frac{\partial P_2}{\partial \theta_2} & \frac{\partial P_2}{\partial \theta_3} & V_3\frac{\partial P_2}{\partial V_3} \\ \frac{\partial P_3}{\partial \theta_2} & \frac{\partial P_3}{\partial \theta_3} & V_3\frac{\partial P_3}{\partial V_3} \\ \frac{\partial Q_3}{\partial \theta_2} & \frac{\partial Q_3}{\partial \theta_3} & V_3\frac{\partial Q_3}{\partial V_3} \end{bmatrix} \begin{bmatrix} \Delta \theta_2 \\ \Delta \theta_3 \\ \Delta V_3/V_3 \end{bmatrix} = \begin{bmatrix} \Delta P_2 \\ \Delta P_3 \\ \Delta Q_3 \end{bmatrix}$$

Starting values. In any iterative method, we must first guess a solution. A "flat start" guess for the unknown voltages is usually $\overline{V}_k = 1 + j0$. Complex voltage values from a previously solved similar network could, instead, be used ("base case start"). A dc power flow can be used to obtain initial angle estimates. The BPA power flow program uses several decoupled power flow iterations before Newton-Raphson iterations. For ill-conditioned cases, good starting values are essential.

Solution algorithm. The basic steps in solving the power flow are:

1. Complex voltages at starting values, convert to rectangular coordinates.
2. Compute the a_{km} and b_{km} variables. Compute the off-diagonal Jacobian terms using Equations B.8–B.11.
3. Based on Equation B.2, sum the a_{km} and b_{km} terms to obtain the current injection at each node. Use Equation B.3 to obtain the calculated active and reactive power injections.
4. Use the power injections of step 3 to compute the diagonal Jacobian terms using Equations B.12–B.15.
5. Use the power injections of step 3 and Equation B.5 to compute the mismatch powers.
6. If all mismatch powers are within tolerance, stop iteration and calculate line flows, losses, and other output quantities. Otherwise go to step 7.
7. Solve Equation B.6 using gaussian elimination methods.
8. Update the complex voltages according to:

$$\begin{bmatrix} \theta \\ \mathbf{V} \end{bmatrix}^{new} = \begin{bmatrix} \theta \\ \mathbf{V} \end{bmatrix}^{old} + \mu \begin{bmatrix} \Delta \theta \\ \Delta \mathbf{V} \end{bmatrix}$$

where μ provides step size control. It is normally equal to unity but is reduced when the power flow is close to non-convergence [4].

9. Convert new voltages to rectangular coordinates. Go to step 2 for next iteration.

The control equations can either be included in the Newton-Raphson iterations or be enforced outside the main iteration loop [3,5].

Example B-3. For the three bus system of Examples B-1 and B-2, compute Newton-Raphson iterations using a flat start. The computations can be done using a spreadsheet or a math CAD program.

Solution: The specified voltage magnitudes for nodes 1 and 2 are 1.05 per unit. The initial guess for the state variables are $\theta_2 = \theta_3 = 0$ and $V_3 = 1.0$. Table B-1 show the mismatch variables and the deviation of the state variables (Example B-2) for the first three iterations. The final result is $\theta_2 = -16.7°$, $\theta_3 = -22.4°$, and $V_3 = 1.00$.

Table B-1

	Iteration 1	Iteration 2	Iteration 3
ΔP_2, per unit	0	-0.1056786884	-0.0001315369
ΔP_3, per unit	-6.0000000000	0.2159529170	0.0000309655
ΔQ_3, per unit	0.3024075017	-0.3032039084	-0.0015803322
$\Delta\theta_2$, radians	-0.2896772684	-0.0026577955	0.0000031588
$\Delta\theta_3$, radians	-0.3892358739	-0.0023005879	0.0000036727
$\Delta V_3/V_3$, per unit	0.0052539766	-0.0051209943	-0.0000272530

Sensitivity analysis. Once convergence is reached, Equation B.6 is a linearized model of the network around the solution point. Changes (sensitivities) of complex voltages to small changes in real and reactive power can be computed. Sensitivities are total derivatives. The inverse of the Jacobian matrix is, in fact, a sensitivity matrix.

Sensitivity analysis can be used for design of voltage control and reactive power compensation. For example, the sensitivity for a 100 MVAr reactive power injection at a candidate bus for a shunt capacitor bank addition can be computed; the voltage increases at nearby busses are obtained. Control options can be specified to represent snapshots in time following an outage. Area interchange, on-load tap changing, or shunt reactive controls can be either active or inactive.

Near the nose of a P–V or Q–V curve, sensitivities get very large and then reverse sign. The Jacobian becomes singular at the maximum power point. This is called a bifurcation point.

Example B-3. For the three bus system of our examples, compute the sensitivity of a 100 MVAr injection at bus 3.

Solution: The 100 MVAr injection increases the voltage magnitude at bus 3 from 1.000 to 1.017. Voltage magnitudes at the other busses do not change since they are fixed.

Other sensitivity formulations are possible to study the effect of controls [6–9]. For example, the change in generator or SVC reactive power for a change in load or shunt compensation may be useful (see example in Chapter 6). Without special programming, various sensitivities can be obtained by running a second case with small changes. For example, the reactive load at a bus or group of busses can be increased by a small amount. The resulting change in bus voltages, line flows and generator reactive power outputs show how busses, branches, and generators "participate." Using starting values from the first case, convergence will be rapid.

For voltage stability, sensitivity analysis must be used with caution. The linearized model is only valid for small changes. An increase in loading may cause generator current limiting (PV bus changed to PQ bus) and drastic changes in sensitivities. As discussed in Chapter 9, extended sensitivity analysis has, however, been developed for voltage stability analysis [7]. The approach includes sensitivity of generator reactive generation for a change in load. The load increase that requires bus type switching (PV to PQ bus) and sensitivity matrix recomputation is thus determined.

For on-line use, the rapid sensitivity computations are preferred over the modal analysis method described next.

B-3 Modal Analysis of Power Flow Model

The modal or eigenvalue analysis method of Gao, Morison, and Kundur [10] is part of the EPRI voltage stability analysis program VSTAB. Modal analysis is akin to sensitivity analysis but the modal separation provides additional insight. VSTAB is a production- grade program for analysis of large systems that has been tested by, and is being used by, several large electric utility companies.

The following is taken from Reference 10. Equation B.6 can be rewritten as:

$$\begin{bmatrix} \Delta \mathbf{P} \\ \Delta \mathbf{Q} \end{bmatrix} = \begin{bmatrix} \mathbf{J}_{P\theta} & \mathbf{J}_{PV} \\ \mathbf{J}_{Q\theta} & \mathbf{J}_{QV} \end{bmatrix} \begin{bmatrix} \Delta \theta \\ \Delta \mathbf{V} \end{bmatrix} \tag{B.16}$$

where the partitioned Jacobian reflects a solved power flow condition and includes enhanced device modeling. By letting $\Delta \mathbf{P} = \mathbf{0}$, we can write:

$$\Delta \mathbf{P} = \mathbf{0} = \mathbf{J}_{P\theta}\Delta\theta + \mathbf{J}_{PV}\Delta\mathbf{V} \text{ or } \Delta\theta = -\mathbf{J}_{P\theta}^{-1}\mathbf{J}_{PV}\Delta\mathbf{V}$$

$$\Delta \mathbf{Q} = [\mathbf{J}_{QV} - \mathbf{J}_{Q\theta}\mathbf{J}_{P\theta}^{-1}\mathbf{J}_{PV}]\Delta \mathbf{V} = \mathbf{J}_R \Delta \mathbf{V} \tag{B.17}$$

where

$$\mathbf{J}_R = [\mathbf{J}_{QV} - \mathbf{J}_{Q\theta}\mathbf{J}_{P\theta}^{-1}\mathbf{J}_{PV}] \tag{B.18}$$

Also,

$$\Delta \mathbf{V} = \mathbf{J}_R^{-1}\Delta \mathbf{Q} \tag{B.19}$$

$\mathbf{J}_R$ is a reduced Jacobian matrix of the system. $\mathbf{J}_R$ directly relates the bus voltage magnitude and bus reactive power injection.

Let λ_i be the ith eigenvalue of $\mathbf{J}_R$ with ξ_i and η_i the corresponding column right eigenvector and row left eigenvector, respectively. The ith modal reactive power variation is

$$\Delta \mathbf{Q}_{mi} = K_i \xi_i \tag{B.20}$$

where

$$K_i^2 \sum_j \xi_{ji}^2 = 1 \tag{B.21}$$

with ξ_{ji} the jth element of ξ_i.

The corresponding ith modal voltage variation is

$$\Delta \mathbf{V}_{mi} = \frac{1}{\lambda_i}\Delta \mathbf{Q}_{mi} \tag{B.22}$$

The magnitude of each eigenvalue λ_i determines the weakness of the corresponding modal voltage. The smaller the magnitude of λ_i the weaker the corresponding modal voltage. If $\lambda_i = 0$, the i-th modal voltage will collapse because any change in that modal reactive power will cause infinite modal voltage variation.

If all the eigenvalues are positive, the system is considered voltage stable. This is different from dynamic systems where eigenvalues with negative real parts are stable. The relationship between system voltage stability and eigenvalues of the $\mathbf{J}_R$ matrix is best understood by relating the eigenvalues with the *V–Q* sensitivities of each bus (which must be positive for stability). $\mathbf{J}_R$ can be taken as a symmetric matrix and therefore the eigenvalues of $\mathbf{J}_R$ are close to being purely real. If all the eigenvalues are positive, $\mathbf{J}_R$ is positive definite and the *V–Q* sensitivities are also positive, indicating that the system is voltage stable.

The system is considered voltage unstable if at least one of the eigenvalues is negative. A zero eigenvalue of $\mathbf{J}_R$ means that the system is on the verge of voltage instability. Furthermore, small eigenvalues of $\mathbf{J}_R$ determine the proximity of the system to being voltage unstable.

The participation factor of bus k to mode i is defined as

$$P_{ki} = \xi_{ki}\eta_{ik} \tag{7.7}$$

For all the small eigenvalues, bus participation factors determine the areas close to voltage instability.

In addition to the bus participations, modal analysis also calculates branch and generator participations. Branch participations indicate which branches are important in the stability of a given mode. This provides insight into possible remedial actions as well as contingencies which may result in loss of voltage stability. Generator participations show which machines must retain reactive reserves to ensure stability of a given mode.

For a practical system with several thousand buses it is impractical and unnecessary to calculate all the eigenvalues. Calculating only the minimum eigenvalue of $\mathbf{J}_R$ is not sufficient because there are usually more than one weak mode associated with different parts of the system, and the mode associated with the minimum eigenvalue may not be the most troublesome mode as the system is stressed. The m smallest eigenvalues of $\mathbf{J}_R$. are the m least stable modes of the system. If the biggest of the m eigenvalues, say mode m, is a strong enough mode, the modes which are not computed can be neglected because they are known to be stronger than mode m.

An Implicit Inverse Lopsided Simultaneous Iteration technique is used to compute the m smallest eigenvalues of $\mathbf{J}_R$ and the associated right and left eigenvectors.

Similar to sensitivity analysis, modal analysis is only valid for the linearized model. Modal analysis can, for example, be applied at points along *P–V* curves or at points in time of a dynamic simulation. References 11 and

12 provide example of modal analysis and comparison with dynamic simulation.

B.4 Fast Decoupled Methods

Decoupled methods take advantage of the physical weak coupling (during normal operation) between active power and voltage magnitude, and reactive power and voltage angle. In the Newton-Raphson method this is reflected in small values for the off-diagonal terms in Equation B.16. We could set the off-diagonal terms to zero and then alternately solve the two sub-problems. Good accuracy is insured by the mismatch equations (Equations B.5 and B.3) which basically means that Kirchoff's laws must be satisfied. Convergence reliability is a key concern.

The decoupled techniques in wide-spread use are the "fast decoupled" methods [13,14]. These methods start with the decoupled Newton-Raphson formulation. Further assumptions lead to constant values for the $\mathbf{J}_{P\theta}$ and $\mathbf{J}_{QV}$ Jacobian matrices. Constant matrices mean that they are computed only once, and that the triangular factorization of gaussian elimination is done only once.

The fast decoupled equations are:

$$\begin{aligned} \Delta \mathbf{P}/\mathbf{V} &= \mathbf{B}'\Delta \theta \\ \Delta \mathbf{Q}/\mathbf{V} &= \mathbf{B}''\Delta \mathbf{V} \end{aligned} \qquad \text{(B.17)}$$

The approximations for the $\mathbf{B}'$ and $\mathbf{B}''$ constant matrices are:

1. Voltages are set to unity.
2. $\cos\theta_{km} = 1$.
3. $G_{km}\sin\theta_{km} = 0$.
4. $Q_k \ll B_{kk}V_k^2$.
5. Terms in $\mathbf{B}'$ that mainly affect reactive power such as shunt reactances are omitted.
6. Phase shifter effects in $\mathbf{B}''$ and possibly $\mathbf{B}'$ are omitted
7. Series resistance in either $\mathbf{B}'$ or $\mathbf{B}''$ is neglected.

The original method [13] neglects series resistance in $\mathbf{B}'$ and is called the XB scheme. The newer method [14] neglects series resistance in $\mathbf{B}''$ and is called the BX scheme. Reference 15 provides the exact equations. The EPRI VSTAB program uses the BX scheme.

Fast decoupled methods have significant speed advantages over the Newton-Raphson method. The limitations, however, are important for voltage stability analysis. Convergence difficulties are experienced in networks with high *R/X* ratios. This could be important if equivalents for subtransmission and distribution circuits are modeled.

Additionally "...the coupled Newton method still has areas of relative strength, such as with very large angles across lines and difficult controls, including special control apparatus that strongly affects MW and MVAr flows" [16]. These limitations are important for the high stress operating conditions of voltage stability analysis. Also in voltage stability analysis, power flow simulation near maximum power points far away from realistic operation is important.

A voltage stability related decoupled formulation. Carpentier [17] developed a decoupled method especially suited to voltage stability analysis. Van Cutsem [18] has applied the method and shows that the method is a good compromise when the reactive power/voltage magnitude subproblem is most important, and when high R/X ratios and high angle differences are present.

B.5 Power Flow Analysis for Voltage Stability

The traditional power flow analysis described above uses PV/PQ bus generator models. As described in Chapter 5, Section 5.1, conversion from PV bus to PQ bus may not adequately represent generator behavior. Instead, the generator should be modeled as a fixed voltage behind an impedance. The fixed voltage depends on controls, including operator actions.

Furthermore, the traditional constant power load models are not always adequate. System representation should be expanded to include subtransmission network and distribution network equivalents, including LTC transformers. Van Cutsem [19] provides an excellent description of power flow models and power flow computation procedures for voltage stability analysis.

B.6 Voltage Stability Static Indices and Research Areas

As already discussed, references 7–10, 17, and 19 describe recent research and development on static analysis of voltage stability. In recent years there has been much additional research. Many of the approaches are described in the proceedings of special conferences on voltage stability held in 1988 and 1991 [20,21]. Only a few of the approaches are outlined here—emphasis is on the most recent work.

A basic requirement is rapid, reliable power flow computation even near the maximum power point. A related requirement is power step size control in computing P–V curves (i.e., smaller step sizes near the nose of the curve). One promising approach is the *continuation* method [22–26], which also allows computation of the underside of P–V curves. Reference 26 shows an example of large-scale power flow computation where a popular conventional power flow program failed convergence some distance from the P–V curve nose computed by the continuation method. In computing *P–V* curves, generation must be properly dispatched.

Indices. At a particular operating point we want an assessment or index of voltage security based on power flow models. For on-line applications, the indices would be part of steady-state security assessment. The indices reflect system robustness to withstand outages or load increases. They should be computationally efficient and easy to understand (even by non-engineer operators). Indices of voltage security should reflect the probability of outages.

CIGRÉ Task Force 38-02-11 categorizes indices as *given operating state based* and *large deviation based*. Large deviation indices account for discontinuities such as generator current limiting. Based on the system's ability to withstand load or power transfer increases, a MW or MVAr distance or margin from the operating point to the maximum power transfer point is determined. This can be done by computing a *P–V* or *Q–V* curve, or by several methods which compute the maximum power transfer capability directly [18,27–29]. Using a more or less conventional power flow program, the maximum power transfer point can be found by a binary search without computing an entire *P–V* curve.

Given operating state indices are based on a solved power flow case or an actual real system operating point. Observation of reactive power reserve is a simple and valuable index that can be calibrated to system security. Observation of voltage magnitudes is also useful, but usually is not as sensitive an index as reactive power reserve. Sensitivity methods based on linearization around an operating point (solved power flow case) are usually repeated at several points along a *P–V* or *Q–V* curve. The indices obtained are more sensitive measures of security than bus voltages. Sensitivity computations such as the total change in generator reactive power for a change in demand [9] is one method (Voltage Collapse Proximity Indicator, *VCPI*). This index changes from near unity at light load to infinity at voltage collapse. The following *VCPI*-based indices may be useful [30]:

$$VCPI_{\mathrm{Pi}} = \frac{\sum \Delta Q_{\mathrm{g}}}{\Delta P_{\mathrm{i}}}$$

$$VCPI_{\mathrm{qi}} = \frac{\sum \Delta Q_{\mathrm{g}}}{\Delta Q_{\mathrm{i}}}$$

where ΔQ_{g} is the change in reactive power output at generators and SVCs for a change in active or reactive load at bus i. The busses with the largest values of $VCPI_{\mathrm{Pi}}$ are the most effective locations for emergency load shedding, and the busses with the largest values of $VCPI_{\mathrm{Qi}}$ are the most effective locations for reactive power compensation.

The eigenvalue or modal analysis method described above [10] is a second method for given operating state indices. Somewhat similar method employ minimum singular value computations [31–33]. Minimum eigenvalues or minimum singular values are also more sensitive indicators of voltage collapse than bus voltages. The indices can be used to determine the megawatt or megavar step size for the next power flow computation.

Besides an index, most of the methods [7–10,18,24,27–33] provide additional insight into effective locations for system reinforcements and into effective operating strategies.

Once an index or a set of indices is chosen, *criteria* for design and operation, and *thresholds* for corrective action are required. There are few industry guidelines at present for criteria and thresholds.

Other research. One area of research centers on defining voltage control areas [34]. These methods would simplify system design and operation.

Other research areas involve optimal power flow techniques and combining power flow methods with artificial intelligence techniques.

The trend is away from purely static traditional models to models and methods incorporating or approximating system dynamics. The maximum power transfer point (nose of a *P–V* curve) is not necessarily the instability point. For example, eigenvalue analysis at a given operating state could be based on dynamic models. Appendix D describes dynamic analysis.

References

1. A. S. Debs, *Modern Power Systems Control and Operation*, Kluwer Academic Publishers, Boston, 1988.
2. J. Arrillaga and C. P. Arnold, *Computer Analysis of Power Systems*, John Wiley & Sons, Chichester, 1990.
3. H. W. Dommel, *Notes on Power Systems Analysis*, University of British Columbia, 1975.
4. A. J. Wood and B. F. Wollenberg, *Power Generation Operation & Control*, John Wiley & Sons, New York, 1984.
5. B. Stott, "Review of Load-Flow Calculation Methods," *Proceedings of the IEEE*, Vol. 62, No. 7, pp. 916–929, July 1974.
6. W. F. Tinney and H. W. Dommel, "Steady-State Sensitivity Analysis, *Proceedings of 4th Power Systems Computation Conference*, Report No. 3.1/10, 1972.
7. N. Flatabø, R. Ognedal, and T. Carlsen, "Voltage Stability Condition in a Power Transmission System Calculated by Sensitivity Methods," *IEEE Transactions on Power Systems*, Vol. 5, No. 4, pp. 1286–1293, November 1990.

8. J. Carpentier, R. Girard, E. Scano, "Voltage Collapse Proximity Indicators Computed from an Optimal Power Flow," *Proceedings of the 8th Power System Computing Conference*, pp. 671–678, August 1984.
9. M. M. Begovic and A. G. Phadke, "Control of Voltage Stability using Sensitivity Analysis," *IEEE Transactions on Power Systems*, Vol. 7, No. 1, pp. 114–123, February 1992.
10. B. Gao, G. K. Morison, and P. Kundur, "Voltage Stability Evaluation Using Modal Analysis," *IEEE Transactions on Power Systems*, Vol. 7, No. 4, pp. 1529–1542, November 1992.
11. G. K. Morison, B. Gao, and P. Kundur, "Voltage Stability Analysis Using Static and Dynamic Approaches," paper 92 SM 590-0 PWRS, presented at the 1992 IEEE/PES Summer Meeting, Seattle, Washington, July 12–16, 1992.
12. CIGRÉ Task Force 38-02-10, *Modelling of Voltage Collapse Including Dynamic Phenomena*, 1993.
13. B. Stott and O. Alsac, "Fast Decoupled Load Flow," *IEEE Transactions on Power Apparatus and Systems*, Vol. 93, pp. 859–869, No. 3, May/June 1974.
14. R. A. M. van Amerongen, "General-Purpose Version of the Fast Decoupled Loadflow," *IEEE Transactions on Power Systems*, Vol. 4, No. 2, pp. 760–770, May 1989.
15. R. Bacher, discussion of Reference 12, *Ibid.*
16. B. Stott and O. Alsac, discussion of Reference 12, *Ibid.*
17. J. L. Carpentier, "'CRIC,' a New Active-Reactive Decoupling Process in Load Flows, Optimal Power Flows and System Control," *Proceedings of the IFAC Conference on Power Systems and Power Plant Control*, Beijing, pp. 59–64, August 1986.
18. T. Van Cutsem, "A Method to Compute Reactive Power Margins with Respect to Voltage Collapse," *IEEE Transactions on Power Systems*, Vol. PWRS-6, No. 2, pp. 145–156, February 1991.
19. T. Van Cutsem, "Voltage Stability : Fast Simulation and Decision Tree Approaches," NERC/EPRI Forum on Operational and Planning Aspects of Voltage Stability, Breckenridge, Colorado, September 14–15, 1992.
20. *Proceedings: Bulk Power System Voltage Phenomena—Voltage Stability and Security*, EPRI EL-6183, January 1989.
21. *Proceedings: Bulk Power System Voltage Phenomena II—Voltage Stability and Security*, Deep Creek Lake, Maryland, 4–7 August 1991. (Includes bibliography of 182 papers written between 1988 and 1991.)
22. R. Seydel, *From Equilibrium to Chaos—Practical Bifurcation and Stability Analysis*, Elsevier Science Publishers, North-Holland, 1988.
23. K. Iba, H. Suzuki, M. Egawa, and T. Watanabe, "Calculation of Critical Loading Conditions with Nose Curve Using Homotopy Continuation Method," *IEEE Transactions on Power Systems*, Vol. 6, No. 2, pp. 584–593, May 1991.
24. V. Ajjarapu and C. Christy, "The Continuation Power Flow: a Tool for Steady State Voltage Stability Analysis," *IEEE Transactions on Power Systems*, Vol. 7, No. 1, pp. 416–423, February 1992.
25. C. Christy, V. Ajjarapu, and B. Srinivasu, "An Approach to Study Steady-State Voltage Stability of Large Scale Power Systems," *Proceedings: Bulk Power System Voltage Phenomena II—Voltage Stability and Security*, Deep Creek Lake, Maryland, pp. 333–340, 4–7 August 1991.

26. R. W. Roddy and C. D. Christy, "Voltage Stability on the MAPP-MAIN Transmission Interface of Wisconsin, NERC/EPRI Forum on Operational and Planning Aspects of Voltage Stability, Breckenridge, Colorado, September 14–15, 1992.
27. F. L. Alvarado, and T. H. Jung, "Direct Detection of Voltage Collapse Conditions," *Proceedings: Bulk Power System Voltage Phenomena—Voltage Stability and Security*, EPRI EL-6183, January 1989.
28. C. A. Cañizares and F. L. Alvarado, "Point of Collapse and Continuation Methods for Large AC/DC Systems," paper 92 WM 103-2 PWRS, IEEE/PES 1992 Winter Meeting, New York, NY, January 26-30, 1992.
29. R. R. Austria, N. D. Reppen, J. A. Uhrin, M. Patel, and A. Galatic, "Applications of the Optimal Power Flow to Analysis of Voltage Collapse Limited Power Transfer," *Proceedings: Bulk Power System Voltage Phenomena II—Voltage Stability and Security*, Deep Creek Lake, Maryland, pp. 311–320, 4–7 August 1991.
30. H. Suzuki, Study Group 37 discussion, *CIGRÉ, Proceedings of the 34th Session*, Vol. II, pp. 37–39, 1990.
31. A. Tiranuchit and R. J. Thomas, "A Posturing Strategy Against Voltage Instabilities in Electric Power Systems," *IEEE Transactions on Power Systems*, Vol. 3, No. 1, pp. 87–93, February 1988.
32. P.-A. Löf, T. Smed, G. Andersson, and D. J. Hill, "Fast Calculation of a Voltage Stability Index," *IEEE Transactions on Power Systems*, Vol. 7, No. 1, pp. 54–64, February 1992.
33. P.-A. Löf, G. Andersson, and D. J. Hill, "Voltage Stability Indices for Stressed Power Systems," paper 92 WM 101-6 PWRS, IEEE/PES 1992 Winter Meeting, New York, NY, January 26-30, 1992.
34. R. A. Schlueter, I-Pung Hu, and Trong-Yih Guo, "Dynamic/Static Voltage Stability Security Criteria," *Proceedings: Bulk Power System Voltage Phenomena II—Voltage Stability and Security*, Deep Creek Lake, Maryland, pp. 265–303, 4–7 August 1991.

Appendix C

Power Flow Simulation Methodology

Although voltage stability is a dynamic phenomenon, simpler and less computationally intense power flow analysis is very valuable. Power flow analysis is especially appropriate if much of the load is non-motor. Power flow simulation has been successfully used for detailed postmortem analysis of actual system events described in Appendix F.

Appendix B describes computer methods for power flow programs.

Following a disturbance, or during a load buildup, we simulate power flow at *snapshots* in time. As shown on Figure 2-1, the response time of equipment ranges from less than a second to many minutes. Steady-state analysis is meaningful only for time frames involving the slower phenomena.

Assume a large disturbance. A loss of generation disturbance causes a large load-generation imbalance. For transmission outage disturbances, the initial reduction in voltage sensitive load also causes a significant load-generation imbalance. Several points in time are suitable for steady-state analysis:

- Ten to thirty seconds following the disturbance. The system should be relatively quiescent with synchronizing swings damped out. On-load tap changer control, overexcitation limiting, and automatic generation control are not significant. Loads are voltage sensitive. The "governor" power flow method described in Chapter 5 should be used.
- Two to five minutes following the disturbance. On-load tap changing may be complete. Tap changer regulation near loads causes restoration of voltage sensitive loads. Field current of generators may be reduced to near rated value. Automatic generation control may be complete if the generation-load imbalance is not large.

- Five minutes or longer following the disturbance. Aggregated loads suffering voltage drop may be restored by thermostatic control. AGC, generation rescheduling and economic dispatch, and operator-initiated procedures may be complete.

The time frames are not sharply separated. We must understand the system and the equipment to make reasonable assumptions. For conventional power flow programs, we suggest procedures for post-disturbance power flow, *V–Q* curves, and *P–V* curves.

Procedure for longer-term voltage stability and governor power flow. We wish to simulate the most onerous moment, i.e., after tap changing and generator current limiting is largely complete, but before slow or manual corrective actions. This moment is often two–three minutes after the disturbance.

Assumptions:

1. For large generator outages, normal automatic generation control (AGC) will not significantly change generation within a two to three minute time frame—either because of being too slow or because of being intentionally suspended after the disturbance. This means power flow program area interchange control is not active. For heavy load conditions, AGC frequency bias coefficients will be close to area natural regulation and Area Control Errors in external control areas will be small.
2. All active turbine-generators speed controls have the same droop setting (typically 5%). Deadband/backlash effects are ignored—some hydro governors have vibrating motors (dither modulation) to reduce deadbands.
3. System frequency is uniform and turbine outputs have reached steady-state values.

Modeling techniques.

1. The voltage sensitive areas of the network are represented in as much detail as feasible. The subtransmission network is represented so that: (a) load and generation are modeled separately (without netting load against generation), (b) subtransmission paths parallel to transmission paths are modeled, and (c) subtransmission reactive power compensation is modeled separately from reactive power load.
2. Loads are modeled as voltage sensitive, with load voltage controlled by LTC transformers and switched shunt capacitor

banks. Constant power loads may be used for more approximate simulation.

3. Generators in the voltage sensitive areas have reactive limits modeled as a function of their active power loading. Preferably, however, the conventional PV/PQ bus generator model is replaced with a detailed model as described in Chapter 5, Section 1. The detailed model depends on the overexcitation control actually installed, and on any expected operator actions.
4. Static var compensators are modeled so they are capacitors at their boost limit. They can be modeled as a thyristor controlled reactor, fixed capacitor (TCR-FC) type of SVC.

Procedure for simulation:

1. Generation-load imbalance should be distributed throughout the interconnection among units with active governors. The imbalance can be distributed in proportion to P_{max} values because we assume that all units have the same droop setting, and see the same frequency deviation. All generators with active governors will pick up their share of generation deficiency until they reach their maximum output. Generators loaded to P_{max} or without active governors must not be included in the distribution of the imbalance. AGC (power flow area interchange control) can be represented if the generation-load imbalance is not large.
2. Additional system losses should also be distributed in proportion to the P_{max} values for all generators on active governor control. Otherwise, the slack bus generation will compensate for all losses. Several runs may be necessary to determine the losses and the P_{max} allocation.
3. The system slack bus should be located far from the area with voltage problems. A nearby slack bus, not having reactive limits, may hold voltages unrealistically high.
4. The effects of transient stability controls which operate immediately following the disturbance are modeled. Controls may include generator tripping, load shedding, sustained fast valving, series capacitor insertion, and shunt capacitor bank switching.
5. Manually controlled or very slow equipment represented in the pre-disturbance power flow as operating automatically should be frozen in the pre-disturbance power flow position—unless there is specific information to do otherwise. Examples include

shunt capacitor banks and reactors, remote (high side) generator voltage control, and LTC transformers (including phase shifters) that are manually controlled. Generator voltage regulator line drop compensation should be represented where used.

6. Acceptability of the results should be based on criteria that measures the system proximity to voltage instability. A minimum voltage requirement is normally not sufficient to determine proximity to voltage instability.
7. Appropriate sensitivity simulations related to the assumptions are advisable.

Procedures for V–Q curves. *V–Q* curves are one of the simplest methods to assess voltage stability. As described in Chapter 2, a fictitious synchronous condenser is applied at a test bus. The test bus becomes a PV bus and a series of voltages are scheduled. The reactive power output versus voltage is plotted. *V–Q* curves can be computed for pre- and post-disturbance conditions and for various load modeling assumptions. Some things to keep in mind are:

- An idea behind *V–Q* (and also *Q–V* curves) is to test system robustness by adding reactive load. Reactive load and voltage performance are closely related. Van Cutsem [1] discusses this concept.
- *V–Q* curves are used to examine reactive power compensation. The test bus shunt reactive compensation characteristics can be superimposed on the *V–Q* curves.
- The meaningful voltage range is about 0.9 to 1.1 per unit. Modeling may be inadequate outside this range.
- Experimentation is required to determine suitable test busses. The best test bus may be depend on the particular disturbance.
- *V–Q* curves are most informative when distinct "voltage control areas" can be found.
- *V–Q* curves may apply undue stress at a single bus. Sensitivity cases can be run using *Q–V* curves where additional reactive load is applied at all load busses in an area.
- Superimposing plots of generator reactive power on *V–Q* curves provides insight into voltage stability.

For large-scale systems, *V–Q* curves are not very realistic and better methods are now available.

Procedures for P–V curves. *P–V* curves (and *S–V* or *Q–V* curves) are a more general method of assessing voltage stability. Load area power can be increased, or transfer across an interface can be increased by generation redispatch. Some concepts are:

- The generation dispatch methods to increase transfer are important and depend on the time frame being simulated.
- At points along *P–V* curves, voltage stability can be assessed by linearized system sensitivity analysis or steady-state modal analysis. These methods, described in Appendix B, provide additional insight.
- For study of system robustness following a large disturbance, the power transfer is increased for the outage condition. The system ability to withstand additional contingencies or load buildup is assessed.
- As load is increased, and if voltage sensitive load is considered, the incremental load should be separated from the base load. The incremental load may be temperature-sensitive air conditioning or heating load.
- For study of load buildup, the power transfer is increased for normal conditions. Distance from the nose of the curve provides voltage stability margin. The ability to withstand contingencies at different load or transfer levels can also be simulated.
- *P–V* curves may be computed for load increase using quasi-dynamic methods. Operation of time-controlled equipment is approximately represented. See Appendix D.

References

1. T. Van Cutsem, "A Method to Compute Reactive Power Margins with Respect to Voltage Collapse," *IEEE Transactions on Power Systems*, Vol. PWRS-6, No. 2, pp. 145–156, February 1991.

Appendix D

Dynamic Analysis Methods for Longer-Term Voltage Stability

Interconnected power systems spanning thousands of kilometers are the largest man-made dynamic systems. Voltage stability adds another dimension and another layer of complexity to these dynamic systems. Dynamic analysis includes time domain simulation and linearized system analysis (eigenanalysis). Computer programs (transient stability programs) for dynamic simulation of transient voltage stability are widely available. In this appendix we discuss simulation of longer-term phenomena.

Careful power flow analysis is often adequate for simulation of the slower forms of voltage stability. Experts have argued that time domain simulation is not needed for longer-term voltage stability analysis. We offer, however, the following reasons for dynamic simulation of minutes to tens of minutes:

1. Time coordination of equipment where the time frames are overlapping; e.g., generator controls, switched capacitor banks, SVCs, and undervoltage load shedding. For example, the timing of reactive power compensation application can affect final results.
2. Clarification of phenomena and prevention of overdesign. Time domain simulation forces more careful analysis and modeling.
3. Confirmation of less computationally intensive static analysis.
4. Improved simulation fidelity especially near stability boundaries.
5. Simulation of fast dynamics associated with the final phases of a collapse.
6. Demonstration and presentation of system performance by easy to understand time domain plots showing time evolution of voltage stability phenomena. More convincing certification of system performance.

7. Education and training.

Modeling. Chapters 3–5 and the companion book *Power System Stability and Control* provide detailed description of equipment modeling. Generator current limiting, network voltage regulation, load control, and energy supply systems (e.g., boilers) must be correctly represented. Controls such as automatic generation control (AGC) must be modeled. Models for slow area load increase may be needed.

Basic models for transient stability are differential and algebraic equations. Controls and relays add logic equations and timers. Digital controls add more complexities such as non-windup integrators or accumulators. For longer-term dynamics, tap changing can be thought of a discontinuous subsystem akin to difference equations [1]. Switching discontinuities may occur frequently in the course of time domain simulation.

Time domain simulation. Several approaches have been taken for efficient long-term simulation of the voltage stability of large systems. The simplest and computationally fastest approach is a sequence of power flow simulations at appropriate snapshots in time. A new power flow is run whenever a switching event, such as transformer tap changing, is required. Load increase over time can be represented.

Lachs [2] has used this method. References 3–6 describe recent use of these quasi-dynamic methods. Appropriate power flow modeling is important and has been incorporated into the recent programs.

The next step in complexity is representing the continuous slow dynamics by differential equation models. Generator and motor flux, and electromechanical dynamics are ignored (represented algebraically), as are generator excitation system dynamics. Uniform system frequency is assumed with a single "swing equation" for total system electrical and mechanical power. Dynamic (differential equation) models are used for prime movers and prime mover controls. Thermostatic control of aggregated loads may be represented by dynamic models. Induction disc type relays (or equivalent integrating electronic relays) used for reactive compensation switching or undervoltage load shedding are modeled dynamically. Discrete actions (tap changing, reactive compensation switching, load shedding) are represented with appropriate time delays once switching criteria is met.

The EPRI LOTDYS program [7] developed in the 1970s had most of these features. The EPRI Operator Training Simulator program [8–10] is also of this type. The Operator Training Simulator is very fast, capable of running relatively large systems in real time. Models for voltage stability analysis are included.

The disadvantage of the uniform frequency method is that fast transients are ignored. One approach is to switch to a conventional transient stability program whenever a major discontinuity such as generator tripping occurs. The uniform frequency program is reinitialized and restarted once the fast transients have damped out. Switching back and forth between two programs, however, is cumbersome.

An approach [11] to unify short and long-term programs was implemented in an early version of the EPRI Extended Transient Mid-Term Stability Program (ETMSP). Artificial damping is added at appropriate times to suppress synchronizing oscillations relative to the center of inertia. The artificial damping can be automatically removed following a switching event. The program uses simultaneous implicit solution algorithms with trapezoidal rule integration [12]. The artificial damping allows time steps to increase from one cycle to as long as one second. Testing of this method was limited to moderately sized systems without induction motor loads.

A more general approach employs variable step size integration methods suitable for "stiff" systems [13]. A stiff system of differential equations has a wide range of time constants. The normal transient stability simulation problem is moderately stiff; the long-term dynamics problem is stiff. Complex algorithms are needed to cope with nonlinearities, and switching events such as load shedding and tap changing.

The EUROSTAG program [14] implements these methods. Integration step size is automatically controlled and may range from 1 millisecond to 100 seconds. The program is valuable for voltage stability analysis [15]. Eigenvalues can be calculated at quasi-dynamic equilibrium points reached during the time simulation.

A different approach is to automatically switch back and forth from dynamic (differential equation) simulation to quasi-steady-state (algebraic) simulation as required by system conditions [3,4].

The EPRI ETMSP 3.0 program [16] features implicit integration and comprehensive models for voltage stability and for short and long term analysis of other phenomena. It is capable of simulating very large systems.

Small disturbance analysis. For additional insight into the voltage instability mechanism, eigenanalysis [17] can be performed at various points along the time evolution of voltage stability. For example, eigenvalues of the linearized system can be computed at a snapshot in time after fast transients have damped out. This period could be similar to the "governor" power flow described in Chapter 5. During the time period of discrete tap changing, eigenvalues can be computed to show, say, the degradation of voltage stability as tap changing continues. Van Cutsem [1] provides an example of this technique. This is an alternative (or a complement) to the steady state modal analysis method described in Appendix B. The steady state modal analysis focuses on the transmission network properties while a dynamic formulation focuses on generator and motor instability mechanisms.

Research areas. Improved software and hardware for simulation of large power systems over extended time periods are being developed, driven in part by the voltage stability challenge. For example, we are all aware of the tremendous advances in general purpose computer hardware such as RISC workstations. Improvements are being seen in other areas such as parallel processing, supercomputers, artificial intelligence, and Macintosh-style graphical interfaces. The need is for power system simulation software to keep up with the "computer revolution."

References

1. T. Van Cutsem, "Voltage Collapse Mechanisms: A Case Study," *Proceedings: Bulk Power System Voltage Phenomena II—Voltage Stability and Security*, Deep Creek Lake, Maryland, pp. 85–101, 4–7 August 1991.
2. W. R. Lachs, "Dynamic Study of an Extreme System Reactive Deficit," *IEEE Transactions on Power Apparatus and Systems*, Vol. PAS-104, No. 9, pp. 2420–2426, September 1985.
3. A. Kurita, H. Okubo, K. Oki, S. Agematsu, D. B. Klapper, N. W. Miller, W. W. Price, J. J. Sanchez-Gasca, K. A. Wirgau, and T. D. Younkins, "Multiple Time-Scale Power System Dynamic Simulation," paper 92 WM 128-9 PWRS, 1992 IEEE/PES Winter Meeting, New York, January 1992.
4. W. W. Price, D. B. Klapper, N. W. Miller, A. Kurita, and H. Okubo, "A Multi-Faceted Approach to Power System Voltage Stability Analysis," *CIGRÉ*, paper 38–205, 1992.
5. N. W. Miller, R D'Aquila, K. M. Jimma, M. T. Sheehan, and G. L. Comegys, "Voltage Stability of the Puget Sound System Under Abnormally Cold Weather Conditions," paper 92 SM 534-8 PWRS, presented at IEEE/PES Summer Meeting, July 12–16, 1992.
6. General Electric Company, *Long Term Power System Dynamics*, EPRI Final Report, Research Project 90-7, April 1974.
7. T. Van Cutsem, "Voltage Stability : Fast Simulation and Decision Tree Approaches," NERC/EPRI Forum on Operational and Planning Aspects of Voltage Stability, Breckenridge, Colorado, September 14–15, 1992.
8. *Operator Training Simulator*, EPRI Final Report EL-7244, May 1991, prepared by EMPROS Systems International.
9. M. Prais, G. Zhang, A. Bose, and D. Curtice, "Operator Training Simulator: Algorithms and Test Results," *IEEE Transactions on Power Systems*, Vol. 4, No. 3, pp. 1154–1159, August 1989.
10. M. Prais, C. Johnson, A. Bose, and D. Curtice, "Operator Training Simulator: Component Models," *IEEE Transactions on Power Systems*, Vol. 4, No. 3, pp. 1160–1166, August 1989.

11. R. J. Frowd, J. C. Giri, and R. Podmore, "Transient Stability and Long-Term Dynamics Unified," *IEEE Transactions on Power Apparatus and Systems*, Vol. PAS-101, No. 10, pp. 3841–3849, October 1982.
12. EPRI Final Report EL-4610, *Extended Transient-Midterm Stability Package: Final Report*, January 1987.
13. G. D. Byrne and A. C. Hindmarsh, "Stiff ODE Solvers: A Review of Current and Coming Attractions," *Journal of Computational Physics*, Vol. 70, pp. 1–62, 1987.
14. M. Stubbe, A. Bihain, J. Deuse, and J. C. Baader, "STAG - A New Unified Software Program for the Study of the Dynamic Behaviour of Electrical Power Systems, *IEEE Transactions on Power Systems*, Vol. 4, No. 1, pp. 129–138, February 1989.
15. J. Deuse and M. Stubbe, "Dynamic Simulation of Voltage Collapses," paper 92 SM 396-2 PWRS, presented at IEEE/PES Summer Meeting, July 12–16, 1992.
16. EPRI, User's Manual—Extended Transient/Midterm Stability Program Package (ETMSP Version 3.0), prepared by Ontario Hydro, June 1992.
17. P. Kundur, G. J. Rogers, D. Y. Wong, L. Wang, and M. G. Lauby, "A Comprehensive Computer Program Package for Small Signal Stability Analysis of Power Systems," *IEEE Transactions on Power Systems*, Vol. 5, No. 4, pp. 1076–1083, November 1990.

Appendix E

Equivalent System 2 Data

This appendix provides complete steady state and data for Equivalent System 2 of Chapter 6 (Sections 6.3 and 6.4).

Steady state data

Figure E-1 shows the system with some of the main data.

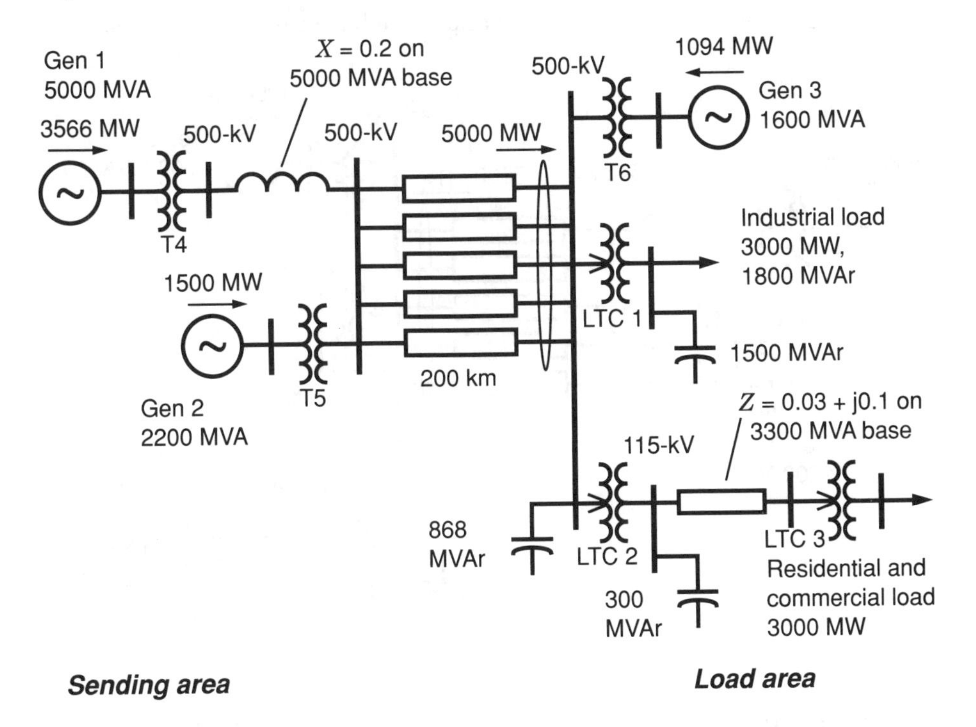

Fig. E-1. Equivalent System 2.

Additional data (on a 100 MVA base) are:

- Transformer LTC 1: 525/13.8-kV, X = 0.003 pu, 500–550 kV tap range, 533 kV base case tap.
- Transformer LTC 2: 525/115-kV, X = 0.003 pu, 500–550 kV tap range, 530 kV base case tap.
- Transformer LTC 3:115/13.8-kV, X = 0.001 pu, 103.5–126.5 kV tap range, 112.1 kV base case tap.
- Transformer T4: 13.2/540-kV, X = 0.002 pu.
- Transformer T5: 13.2/540-kV, X = 0.0045 pu.
- Transformer T6: 13.2/530-kV, X = 0.00625 pu.
- Gen 1: slack bus, V = 0.98 pu.
- Gen 2: V = 0.964 pu, reactive power limits are -200 MVAr and 725 MVAr.
- Gen 3: V = 0.972 pu, reactive power limits are -200 MVAr and 700 MVAr.
- 500-kV transmission lines: Z = 0.0015 + j0.02880, $B/2$ = 1.173.
- Shunt capacitor reactive power on Figure E-1 are at nominal voltage.

Figure E-2 shows base case voltages and some base case power flows.

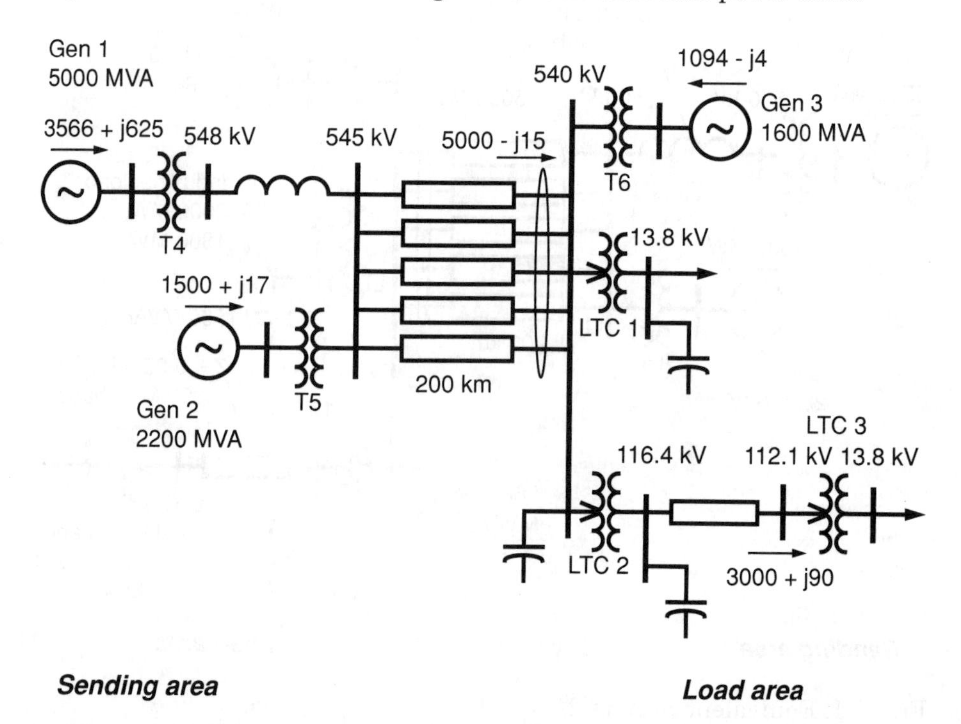

Fig. E-2. Equivalent System 2, base case conditions.

Dynamic data

Generator equivalents. The electrical characteristics of the three generators are identical and correspond to a 0.95 power factor, 590-MVA coal-fired unit. The generator data is based on Unit *F*18 in Appendix D of Anderson and Fouad [1]. The reference describes the generator, excitation system, and governor-turbine models. The generators have a static excitation system with power system stabilizer. The MVA ratings are given on Figures E-1 and E-2. Gen 1 is an equivalent for a large remote system, but is relatively small electrically (5000 MVA) to provide only limited voltage support for the load area.

The inertia constant, H, for Gen 2 and Gen 3 is 2.32 MW-s/MVA. The inertia constant of Gen 1 is very large, modeling the mechanical equivalent of a large interconnection.

Load equivalents. The industrial load is 100% motor. Two equivalent motors are represented. A 3375 MVA motor consumes 2700 MW. Data corresponds to the large industrial motor of Table 4-1. A 500 MVA motor consumes 300 MW. Data corresponds to the small industrial motor of Table 4-1.

The 3000 MW residential and commercial load is half resistive and half motor. The resistive load may be thermostatically controlled. The single motor equivalent is an aggregate of motors heavily dominated by air conditioning load [2, page 3-21]. Motor data is:

Table E-1

MVA	R_s	X_s	X_m	R_r	X_r	A	B	H	LF_m
2440	.056	.092	2.14	.059	.075	.32	0	.342	.615

References

1. P. M. Anderson and A. A. Fouad, *Power System Control and Stability,* The Iowa State University Press, Ames, Iowa, 1977.
2. General Electric Company, *Load Modeling for Power Flow and Transient Stability Computer Studies, Volume 1: Summary Report*, EPRI Final Report EL-5003, January 1987.

Appendix F

Voltage Instability Incidents

In recent years, many voltage instability incidents have occurred around the world. There have also been some close calls, many of which have not been widely reported. We will group known incidents by the time frames defined in Chapter 2—transient and longer term.

Table F-1 lists incidents resulting in collapse. Table F-2 lists incidents not resulting in collapse. Some of the incidents are complex and involve other phenomena besides pure voltage instability. We describe the incidents briefly, first for the Table F-1 list and then for the Table F-2 list. The references listed in the tables provide further description. Many of the best known incidents are described in detail (with plots) in an IEEE Working Group report [1].

Nelson River HVDC system, Winnipeg, Canada, April 13, 1986. Partial voltage collapse occurred during energization of a converter transformer. Inrush current depressed ac voltage, resulting in commutation failures and inverter firing angle advance. Voltage collapsed to 57% and then recovered after temporary dc blocking. A second voltage collapse followed. Tie line tripping, shutdown of three of four dc poles, and undervoltage load shedding resulted. A control (System Undervoltage Protection) which reduces dc power by a fixed amount for low ac voltage was out of service.

SE Brazil, Paraguay, November 30, 1986. After several ac system outages, the São Roque inverter (Itaipu HVDC link) ac voltage dropped to 0.85 pu for several seconds. Repetitive commutation failures occurred, and dc power control increased dc current which increased converter reactive consumption. A complete dc system shutdown and an ac system breakup resulted. Over 1200 MW of load was shed. This and other disturbances led to a number of dc control changes.

South Florida, May 17, 1985. A brush fire caused three lightly-loaded 500-kV lines to trip, resulting in voltage collapse and blackout within a few seconds. Low voltage prevented underfrequency relays from operating. Transient stability simulation indicated the system should have recovered and load modeling deficiencies (including modeling of power plant auxiliaries) are suspected. Load loss was 4,292 MW. Figure 2-3 shows voltage and frequency plots.

Table F-1, Incidents with Collapse

Date	Location	Time Frame	Ref.
13 April 1986	Winnipeg, Canada Nelson River HVDC link	Transient, 1 sec.	2,3
30 Nov. 1986	SE Brazil, Paraguay Itaipu HVDC link	Transient, 2 sec.	4,5
17 May 1985	South Florida, USA	Transient, 4 sec.	1,6
22 Aug. 1987	Western Tennessee, USA	Transient, 10 sec.	7,8,9
27 Dec. 1983	Sweden	Longer term, 55 sec.	1,10,11
2 Sept. 1982	Florida, USA	Longer term, 1–3 min.	12,13
26 Nov. 1982	Florida, USA	Longer term, 1–3 min.	12,13
28 Dec. 1982	Florida, USA	Longer term, 1–3 min.	1,12,13
30 Dec. 1982	Florida, USA	Longer term, 1–3 min.	12,13
22 Sept. 1977	Jacksonville, Florida	Longer term, few min.	14
4 Aug. 1982	Belgium	Longer term, 4.5 min.	15
12 Jan. 1987	Western France	Longer term, 6–7 min.	1,16
9 Dec. 1965	Brittany, France	Longer term	17
10 Nov. 1976	Brittany, France	Longer term	17
23 July 1987	Tokyo, Japan	Longer term, 20 min.	1,18
19 Dec. 1978	France	Longer term, 26 min.	1,19,20
22 Aug. 1970	Japan	Longer term, 30 min.	21

Western Tennessee, August 22, 1987. A 78-cycle, phase-to-phase arcing 115-kV bus fault in Memphis, Tennessee resulted in 161-kV and 500-kV system voltages between 75% and 82% of normal for approximately 10 seconds following fault clearing. Motor load reactive requirements prolonged the voltage depression. Zone 3 relays operated and cascading resulted. Load loss was 1265 MW.

Sweden, December 27, 1983. A disconnector failure and fault at a substation west of Stockholm resulted in loss of the substation and two 400-kV lines. Approximately 8 seconds later, a 220-kV line tripped on overload. On-load tap changer actions caused lower transmission voltages and higher currents on remaining north to south lines. Approximately 50 seconds after the fault, another 400-kV line tripped. Cascading and islanding of southern Sweden followed. Frequency and voltage collapsed and underfrequency load shedding did not save the system. Nuclear units in the islanded system tripped by overcurrent or underimpedance protection resulting in a blackout. Load loss was about 11,400 MW.

Florida, 1982. All four disturbances were similar and were initiated by loss of a large generator unit in central or southern Florida. Because of the increased imports, voltages deteriorated and separations occurred after one to three minutes. The islandings were followed by underfrequency load shedding of about 2000 MW. These disturbances led to implementation of shunt reactor and shunt capacitor bank switching by voltage relays at several 230-kV substations.

Jacksonville, Florida, September 22, 1977. A series of voltage collapses occurred. The collapses involved unit trips, field current limiting, manual load shedding, and other phenomena.

Belgium, August 4, 1982. The initial disturbance was disconnection of a 700 MW unit during commissioning tests. After 45 seconds, automatic control action reduced reactive power on two other units. Three to four minutes after the initial event, three units were tripped by "max MVAr protection." At 3 minutes, 20 seconds, voltage was 0.82 pu at a major generating station. At 4 minutes, 30 seconds, two additional generators tripped by impedance relays, resulting in collapse.

Western France, January 12, 1987. Over a period of about fifty minutes, four units at the Cordemais thermal plant tripped. Voltages decayed and nine other thermal units tripped over the next seven minutes; eight units because of field current protection defects. The total power deficiency was around 9000 MW. Voltages then stabilized at very low levels (0.5–0.8 per unit). After about six minutes of voltage collapse, 400/225-kV transformers were tripped to shed about 1500 MW of load. Voltages then recovered.

The fact that voltages stabilized at low levels rather than completely collapsed is very interesting. The incident occurred in wintertime when the

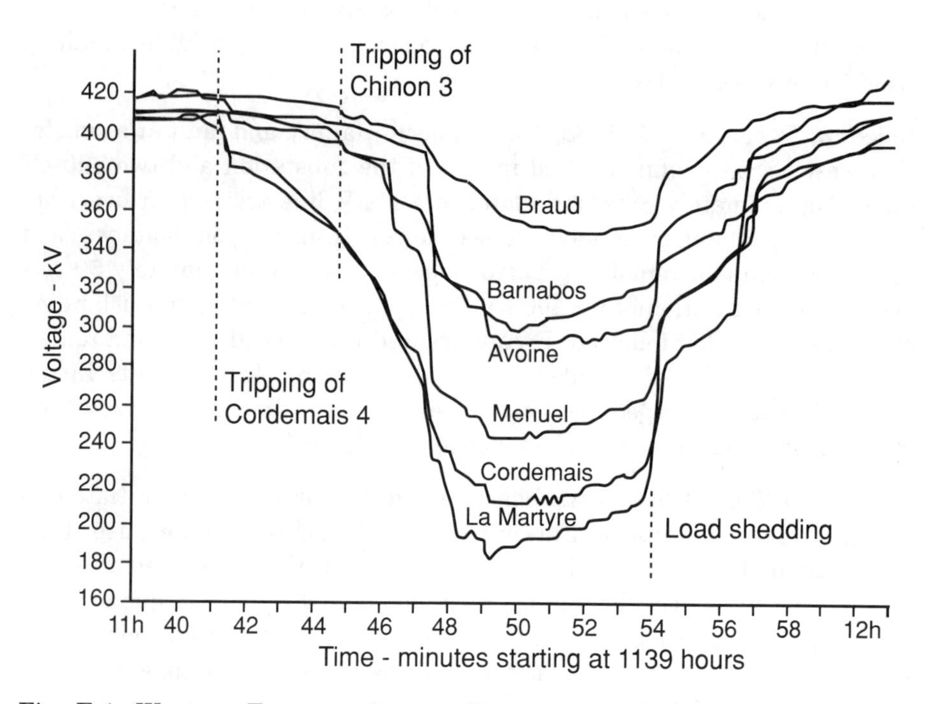

Fig. F-1. Western France voltage collapse on 12 January 1987. Voltage recordings on 400-kV system.

loads were quite voltage sensitive. Some motor load no doubt dropped off line so that the remaining load was even more voltage sensitive. During the low voltage period, adding more load (perhaps by thermostatic control) would probably have resulted in less load power—equivalent to being on the bottom side of a P–V curve. Figure F-1 shows the voltage profiles.

Tokyo, July 23, 1987. Weather was very hot and loads were abnormally high. After the noon hour, loads increased at 400 MW/minute. Despite connection of all available shunt capacitors, the voltage decayed, with voltages on the 500-kV system at 460 kV at 1315 hours and at 370 kV at 1319 hours. Collapse began at 1319 hours; 8,168 MW were interrupted. Unfavorable characteristics of new type of air conditioners were thought to be part of the problem.

France, December 19, 1978. France was importing power from other countries. Load rise between 0700 and 0800 hours was 4,600 MW, compared to 3,000 MW the previous days. Voltage deteriorated after 0800 hours and between 0805 and 0810 some EHV/HV tap changers were blocked. Low voltages reduced some thermal production. At 0820, voltages on the eastern 400-kV system ranged from 342 kV to 374 kV. Cascading began at 0826 after an overload relay tripped a major 400-kV line (system operators had an alarm that the line would trip with 20 minutes time delay). During restoration another collapse occurred. System restoration was completed by 1230. Load interrupted was 29 GW and 100 GWh. The cost of the outage was estimated at $200–300 million.

Table F-2, Incidents without Collapse

Date	Location	Time Frame	Ref.
17, 20, 21 May 1986	Miles City, Montana, USA HVDC link	Transient, 1–2 sec.	22
11, 30, 31 July 1987	Mississippi, USA	Transient, 1–2 sec.	23
11 July 1989	South Carolina, USA	Unknown	24
21 May 1983	Northern California, USA	Longer term, 2 minutes	1,25
10 Aug. 1981	Longview, Wash., USA	Longer term, minutes	26,27
17 Sept. 1981	Central Oregon, USA	Longer term, minutes	28
20 May 1986	England	Longer term, 5 minutes	29,30
2 Mar. 1979	Zealand, Denmark	Longer term, 15 min.	30
3 Feb. 1990	Western France	Longer term, minutes	31
Nov. 1990	Western France	Longer term, minutes	31
22 Sept. 1970	New York state, USA	Longer term, minutes, insecure for hours	1,32
20 July 1987	Illinois and Indiana, USA	Longer term, insecure for hours	33,34, 35,36
11 June 1984	Northeast USA	Longer term, insecure for hours	37
5 July 1990	Baltimore, Washington D.C., USA	Longer term, insecure for hours	38

Miles City HVDC link, May and July 1986. As a result of weak ac systems, large voltage deviations occurred during dc ramping and during reactive switching resulting in commutation failures, and in some cases, converter tripping and loss of a 310 MW west-side generator.

Mississippi, July 1987. In 1981, undervoltage load shedding was installed in a load area where loss of a 500/161-kV transformer source may cause voltage collapse. Air conditioning load comprises a large portion of summer peak load. On three separate days in July 1987, current transformer failures caused transformer bank outage and other outages. Voltages collapsed immediately, but undervoltage load shedding operating within two seconds to shed up to 400 MW of load, resulting in system recovery.

On July 22, 1992, loss of the 500/161-kV transformer resulted in 586 MW of undervoltage load shedding.

South Carolina, July 11, 1989. A nuclear plant generating 868 MW and 440 MVAr tripped during record power demand. Due to automatic voltage regulator action, nine hydro units generating a total of 649 MW then tripped by generator backup relays. The 115-kV transmission voltage dropped to about 89% and the 230-kV voltage dropped to about 93%.

Northern California, May 21, 1983. Following a Pacific HVDC Intertie bipolar outage (1286 MW), voltages along Pacific 500-kV AC Intertie decayed for two minutes. The lowest voltage was 385-kV at Vaca-Dixon 500-kV substation (73% of the 525 kV normal operating voltage). The low voltage caused tripping of pumps at various aqueduct stations, leading to recovery. The initial Pacific AC Intertie loading was 2240 MW.

Longview, Washington area, August 10, 1981. Temperature was very hot (41°C). The Allston 500/230-kV autotransformer near the Trojan nuclear plant was out for maintenance. The 1100 MW Trojan plant tripped, removing power and voltage support to the Longview area. Transmission lines (230-kV and 115-kV) were overloaded and a number of single line to ground faults occurred, probably because of sagging into trees (sagging due to temperature, and current overload due to high load and low voltage). The Longview aluminum reduction plant 13.8-kV voltage dropped as low as 12.4 kV and Bonneville Power Administration operators permitted plant operators to operate tap changers on 230/13.8-kV transformers —this was the wrong thing to do—the voltage rose to 13.2 kV, but then dropped again below 13 kV. Shortly thereafter a potline relayed off. At one point 230-kV system voltage was down to 208 kV and voltage collapse of the Longview area was imminent. Forty-six minutes after the Trojan outage, operators tripped 110 MW of potline load. Lines were then reclosed and the Allston transformer was restored to service.

Central Oregon, September 17, 1981. The 51 MVAr, 230-kV capacitor bank at LaPine Substation began hunting (cycling) following opening of the 230-kV line serving the Bend, Oregon load area from the north. With only the 230-kV line from the south in service, the capacitor bank induction disc voltage relay settings were such that the capacitor bank switched nineteen times in about one hour—at about three minute intervals. The voltage extremes were about 219 and 251 kV. Figure F-2 shows substation voltage. It appears that tap changer regulation of load contributed to the hunting. Following capacitor bank energization voltage increased for about three minutes until bank deenergization; with the bank off, voltage decayed until the bank energization.

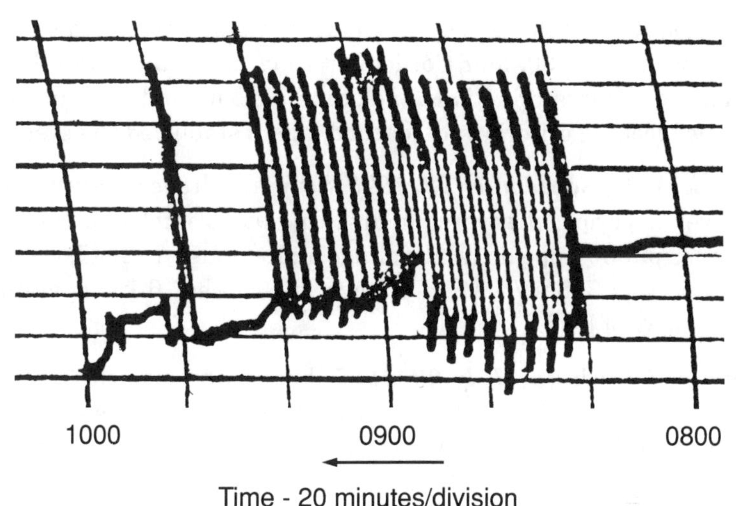

Fig. F-2. Shunt capacitor bank hunting at LaPine Substation near Bend Oregon on September 17, 1981.

England, May 20, 1986. During a thunderstorm, six 400-kV circuits were lost within one minute. Voltage decayed progressively over a five minute period, with 352 kV recorded at the lowest point. Within 5 minutes, 1000 MW of gas turbines were brought on line to stabilize voltages. Circuit reclosure then restored voltage. Predictions were that voltage collapse should have occurred—speculation is that interaction of tap changers with different timer settings slowed the voltage decay, allowing time for operator action.

Zealand, Denmark, March 2, 1979, The initiating event was tripping of a 270 MW unit in the southern part of the island. There were no nearby reactive power reserves and voltage decayed due to load restoration by tap changing. After fifteen minutes the voltage was below 0.75 per unit making it impossible to start and synchronize a 70 MW gas turbine in the area. Load was then shed manually to restore voltage and allow gas turbine synchronization.

The long time for voltage decay (fifteen minutes) was due to relatively long fixed time delay between tap steps.

Western France, February 3, 1990 and November 1990. Measures taken by Electricité de France since the January 12, 1987 incident (Table F-1) contained two very severe disturbances. A violent storm on February 3, 1990 caused loss of the 225-kV and 400-kV buses at the Cordemais generating station. Automatic tap changer blocking and operator network unloading stabilized the power system until repairs were made.

In November 1990, four Cordemais generating units were lost within forty minutes. Operator actions, including blocking of tap changers before automatic blocking criteria was reached, again stabilized the system.

New York state, September 22, 1970. Several voltage decays were experienced over a period of several hours. Operators implemented voluntary reductions, public appeals, and voltage reductions. At 1545, voltage at a 345-kV bus was 318 kV—when voltage fell another 6 kV, operators shed about 200 MW of load.

Illinois and Indiana, July 20, 1987. Loads were at near-record levels, reactive demand was higher than expected, power transfers were high, and many generators were unavailable. Voltages were as much as 8% below normal at 765-kV, 11% at 345-kV, and 12% at 138-kV. As a result, AEP added 138-kV mechanically switched capacitors and put switches on 765-kV shunt reactors. Studies indicated that the system could have survived a credible single contingency.

Northeast United States, June 11, 1984. Cause was combination of abnormally high loads, planned outages, and forced outages. Despite voltage reductions and use of available shunt capacitors, the west-to-east transfers in the Pennsylvania-New Jersey-Maryland Interconnection (PJM) and the New York Power Pool imports from Canada had to be reduced to keep within voltage and reactive precontingency limits. PJM reduced generation on several units to increase reactive power output; replacement power was purchased from gas turbine generators in Virginia.

Baltimore and Washington D.C., July 5, 1990. High loads (high temperatures) and generation outages resulted in low 500-kV voltages.

Voltage reductions (5%), running out-of merit generation, demand-side management, and rotating blackouts (400 MW) were required.

References

1. IEEE Committee Report, *Voltage Stability of Power Systems: Concepts, Analytical Tools, and Industry Experience*, IEEE publication 90TH0358-2-PWR.
2. J. Chand and D. Tang, "Operational Experience of the Nelson River HVDC System," *Proceedings of Second HVDC System Operating Conference*, pp. 21–31, Winnipeg, Canada, 18–21 September 1989.
3. M. M. Rashwan, G. B. Mazur, M. A. Weekes, and D. P. Brandt, "AC/DC System Power/Voltage Stability Enhancement Using Modified Power Control," *Proceedings of Second HVDC System Operating Conference*, pp. 35–39, Winnipeg, Canada, 18–21 September 1989.
4. IEEE Committee Report, "HVDC Controls for System Dynamic Performance," *IEEE Transactions on Power Systems*, Vol. 6, pp. 743–752, May 1991.
5. H. Arakaki, J. C. Lopes, and A. A. S. Praca, "Itaipu HVDC Transmission System—Analysis of Control System and Protection Performance after Two Years of Operation," paper SP-25, *Proceedings of 1st Symposium of Specialists in Electric Operational Planning*, Rio de Janeiro, August 17–21, 1987.
6. D. McInnis, "South Florida Blackout," unpublished Florida Power & Light report.
7. North American Electric Reliability Council, *1987 System Disturbances*, p. 19, July 1988.
8. G. C. Bullock, "Cascading Voltage Collapse in West Tennessee, August 22, 1987," *Georgia Institute of Technology 44th Annual Protective Relaying Conference*, May 2–4, 1990.
9. G. C. Bullock, "Cascading Voltage Collapse in West Tennessee, August 22, 1987," *Western Protective Relaying Conference*, October 1990.
10. CIGRÉ Working Group 38.01, "Planning Against Voltage Collapse," *Electra*, pp. 55–75, March 1987.
11. J. F. Christensen, Study Group 38 discussion, *CIGRÉ, Proceedings of 33rd Session*, Vol. II, 1988.
12. North American Electric Reliability Council, *Review of Selected Major Bulk Power System Disturbances in North America—1982.*
13. IEEE Committee Report, "VAR Management—Problem Recognition and Control," *IEEE Transactions on Power Apparatus and Systems*, Vol. PAS-103, No. 8, pp. 2108–2116, August 1984.
14. W. R. Lachs and D. Sutanto, "Different Types of Voltage Instability," IEEE/PES paper 93 SM 518-1 PWRS.
15. A. J. Calvaer and E. Van Geert, "Quasi Steady State Synchronous Machine Linearization Around an Operating Point and Application," *IEEE Transactions on Power Apparatus and Systems*, Vol. PAS-103, No. 6, pp. 1466–1479, June 1984.
16. Y. Harmand, M. Trotignon, J. F. Lesigne, J. M. Tesseron, C. Lemaitre, and F. Bourgin, "Analysis of a Voltage Collapse Incident and Proposal for a Time-Based Hierarchical Containment Scheme," *CIGRÉ,* paper 38/39-02, 1990.
17. C. Barbier and J-P Barret, "Analysis of Phenomena of Voltage Collapse on a Transmission System," *Revue Generale de l'electricite*, Vol. 89, pp. 672-690, October 1980.

18. A. Kurita and T. Sakurai, "The Power System Failure on July 23, 1987 in Tokyo," *Proceedings of the 27th Conference on Decision and Control*, Austin, Texas, December 1988.
19. A. Cheimanoff and C. Curroyer, "The Power Failure of December 19, 1978," *Revue Generale de l'electricite*, Vol. 89, pp. 280–320, April 1980.
20. J. A. Casazza, *Interim Report on the French Blackout of December 19, 1978*, DOE report, February 8, 1979.
21. T. Nagao, "Voltage Collapse at Load Ends of Power Systems," *Electrical Engineering in Japan*, Vol. 95, No. 4, 1975.
22. R. D. Doherty, R. K. Johnson, S. F. Schweitzer, and T. L. Weaver, "Miles City Station—Early Operating Experience," *CIGRÉ, Proceedings of 32nd Session*, Vol. I, paper 14-03, 1986.
23. H. M. Shuh and J. R. Cowan, "Undervoltage Load Shedding An Ultimate Application for Voltage Collapse," *Georgia Institute of Technology 46th Annual Protective Relaying Conference*, April 29–May 1, 1992.
24. North American Electric Reliability Council, *1989 System Disturbances*, p. 18, July 1990.
25. R. H. Vierra, "Reactive Power and Voltage Control in Western Systems," presented at IEEE/PES summer meeting, Vancouver July 16, 1985.
26. T. R. Reitman, "Longview Disturbance—August 10, 1981," BPA memorandum, August 27, 1981.
27. D. Mann, "BPA Requested Power Curtailment," Reynolds Aluminum memorandum, August 21, 1981.
28. R. Demaris, "LaPine Substation Capacitor Cycling," Bonneville Power Administration memorandum, 22 January 1982.
29. M. G. Dwek, Study Group 38 discussion, *CIGRÉ, Proceedings of 33rd Session*, Vol. II, 1988.CIGRÉ TF 38-02-10, *Modelling of Voltage Collapse Including Dynamic Phenomena*, 1992.
30. CIGRÉ TF 38-02-10, *Modelling of Voltage Collapse Including Dynamic Phenomena*, 1993.
31. B. Heilbronn and G. Testud, Study Group 38 discussion, *CIGRÉ, Proceedings of the 1992 Session* (in French).
32. General Electric Company, *Long-Term Power System Dynamics*, Phase III, EPRI Final Report EL-983, May 1982.
33. North American Electric Reliability Council, *1987 System Disturbances*, p. 18, July 1988.
34. P. B. Johnson, S. L. Ridenbaugh, R. D. Bednarz, and K. G. Henry, "Maximizing the Reactive Capability of AEP Generating Units," *Proceedings of the American Power Conference*, Vol. 52, pp. 373–377, 1990.
35. M. Ea. Rahman and M. D. Higgins, "Reactive Power Planning and its Impact—An Actual Case History," *Proceedings of the American Power Conference*, Vol. 53, pp. 907–911, 1991.
36. J. A. Pinnello, "Planning Perspectives of Voltage Control," *Proceedings: Bulk Power System Voltage Phenomena—Voltage Stability and Security*, EPRI EL-6183, pp. 3-57–3-60, January 1989.
37. North American Electric Reliability Council, *1984 System Disturbances*, p. 24.
38. North American Electric Reliability Council, *1990 System Disturbances*, p. 33–36, May 1991.

Index

ABOUT THE AUTHOR

Carson W. Taylor is Principal Engineer with the Bonneville Power Administration in Portland, Oregon. His areas of expertise at BPA include power system control and protection, system dynamic performance, ac/dc power system interaction, and power system planning. In 1986, he established Carson Taylor Seminars, a company specializing in electric power system education. Mr. Taylor is a Fellow of the IEEE, and the author or coauthor of many IEEE and CIGRE papers.